VISUAL GUIDE TO
BLACKSMITHING

standard publications, inc.

Visual Guide to
Blacksmithing

Susan Bove

Illustrator
Mirko Jugurdzija

Editor
Laura Neff

standard publications, inc.

Cover design: **Mark McCloud**
Layout editor: **Mirko Jugurdzija**
Production editor: **Mano Kime**
Manufacturing producer: **Standard Publications, Inc.**

Copyright ©2009 by Standard Publications, Inc.
Standard Publications, Inc.
Champaign, Illinois 61825

All rights reserved. No part of this book may be reproduced, stored in a retrieval system, or transmitted, in any form or by any means, electronic, mechanical, photocopying, recording, or otherwise, without prior written permission from the publisher.

The author and publisher of this book make no warranty of any kind, express or implied. The author or publisher shall not be held liable in any event for incidental or consequential damages in connecton with, or arising out of, the furnishing, performance, or use of this information.

The publisher offers discounts on this book when ordered in bulk quantities.

`ISBN Number 0-9722691-9-3`

Printed in the United States of America

Standard Publications, Incorporated.

www.standardpublications.com

Contents

1	**Introduction**	**13**
2	**A Word (or Two) about Safety**	**19**
3	**The Shop**	**17**
4	**Equipment**	**37**
	The Forge and Fire	37
	The Anvil	46
	The Quenching Bath	52
	The Coal Box	53
	The Tool Table and Workbench	53
	The Drill Press	56
	The Grinding Wheel	57
	The Abrasive Cut-off Wheel	58
	The Cotton Buffer	58
5	**Essential First Tools**	**63**
	Hammers	63
	Hardies and Sets	66
	Fire Tools	69
	Brushes	72
	Tongs	73
	Measurment Tools	78
6	**Choosing Your Metal**	**85**
	Cast Iron	87
	Wrought Iron	87
	Alloy Steels	88
	So, what exactly is it?	89
7	**Techinques for the Beginner Smith**	**95**
	Hammering	96

Cutting	99
Bending	103
Drawing Out	106
Upsetting	112
Twisting	115
Welding	117
Heat Treatment	121
Case Hardening	126

8 Making Your First Tools — 133

Consider This	133
The Simple Screwdriver	136
The Hammer... A Brief History	141
Designing Your Hammer	141
Hammer Weight	142
Hammer Face	144
Making Your Hammer	146
Bringing it Together with Heads and Rivets	155
The Versatile Tongs	157
Consider This	158
Light Weight Tongs	159
Heavy Weight Tongs	163
Specialty Tongs	167

9 Getting a Handle on Things — 173

Consider This	173
Wooden Handles	174
Power Handles	177
Plastic Handles	181

10 Making Hardware with Personality — 185

Hook and Eye	185
Hasp and Eye	191
Sliding Bolt	194
Hinges	200

11 Decorative, yet Functional, Smithing — 211

- A Place to Hang Your Hat — 211
- Wall Brackets — 215
- Scrolls — 220

12 Finishing Your Creation — 231

- Removing Scale — 232
- Protecting Your Creations — 233
- Painting — 234
- A Cool Treatment — 235
- In Closing — 235

G Glossary — 236

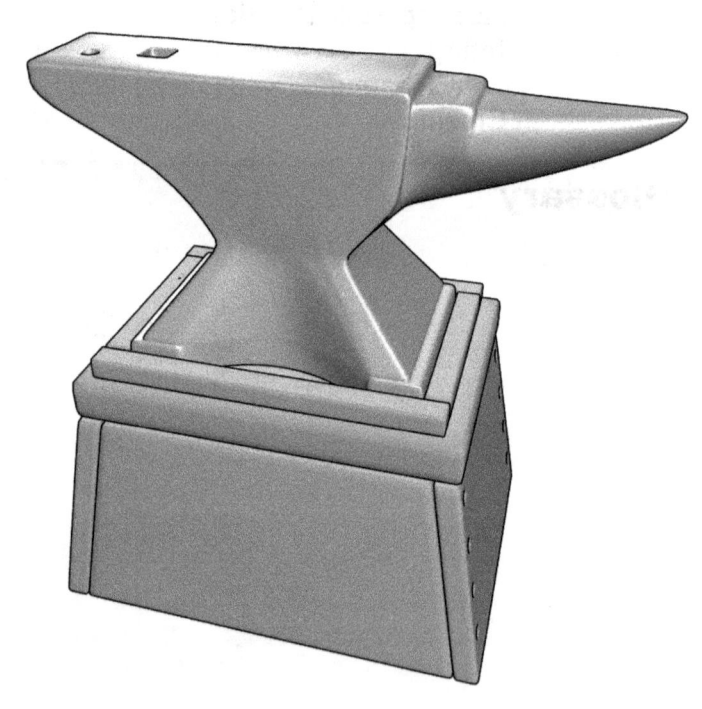

1
Introduction

Introduction

So, you want to be a blacksmith? Good for you! You'll be joining an ancient fraternity of men (and women) who can successfully harness the power of fire and transform metal to their bidding. Men first discovered that metals could be changed when heated back in the Stone Age. The first available materials were impure and comprised of copper and other trace metals, which we now know as bronze. This was a relatively soft metal, but served ancient man better and was more versitile than their old stone weapons and tools. It wasn't until the Bronze Age than men discovered how to separate ore from iron - and that was when modern smithing was born.

The Egyptians were the first people credited with tempering and forging iron with a hammer. When the Egyptian empire crumbled, the technique spread throughout what is now known as Europe and the Middle East. Civilization slowly crept into the Iron Age. Aside from the obvious benefits of using iron to make tools and weapons of war, it was the combination of metal and wood that made the clearing of land for cultivation and wheeled vehicles possible. Considering all the wonderful and practical items that could be made from metal, the smith became a very important member of every community.

You may have noticed that this chapter refers to the craftsman as a "smith" rather than a "blacksmith." Smiths, in the very early years of the craft, worked with different types of metals, eventually, developing a specialty. For example, a smith, whose specialty was working with lead, was referred to as a *whitesmith,* while a smith of iron became known as a *blacksmith.* When you use the term *blacksmith* today, the first thing that comes to mind for most people is a person who makes horseshoes. However, that craftsman is a actually a *farrier.* There were other specialty smiths who made just one product, such as a *chainsmith,* who forged links for a chain and a *nailsmith* (often a woman), who made nothing but nails.

The smith enjoyed his status in the community and was never short of work - until the Industrial Revolution. Factory output of products that were once hand crafted by the smith quickly became widely available. The smith experienced a decreased demand for his services and had to adapt to this change. Some could still find a place for their skills within the industry, while others became factory workers.

There are still craftsmen who can make a living from smithing. However, many who smith do so because they hold a pure love of the craft. Today, blacksmiths are not only sought by necesscity, but also for the artistry of their creations. Whether you are planning on smithing for profit or as a hobby, you will attain tremendous satisfaction from being able to construct a creation through your own hard work and creativity.

Introduction

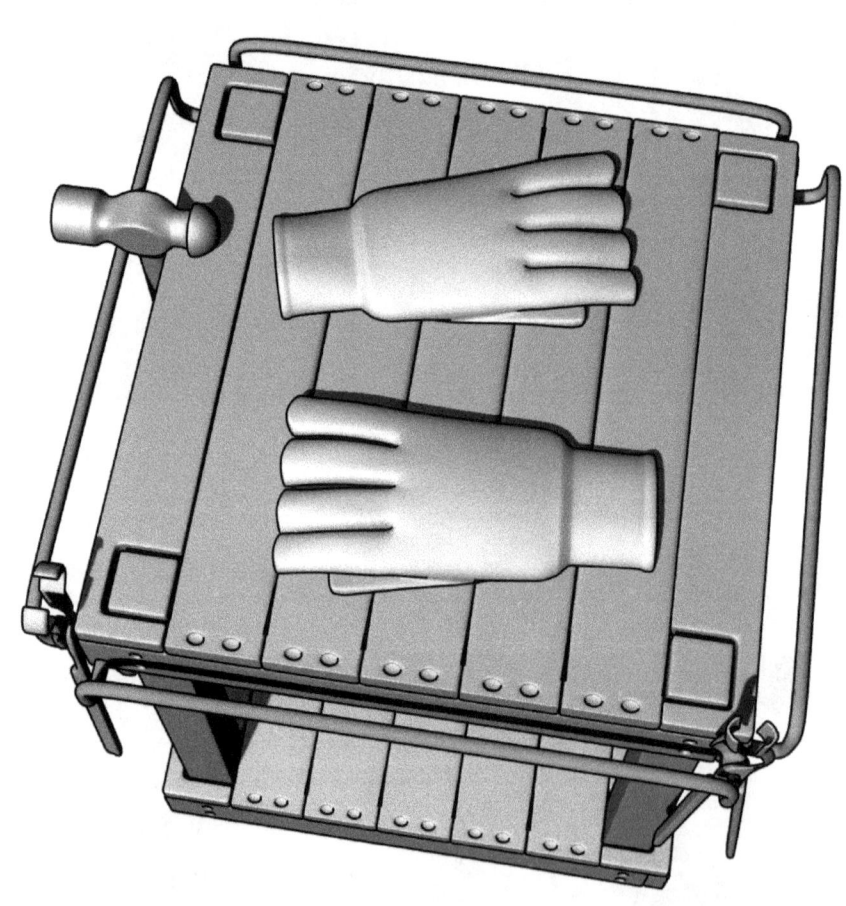

2

A Word (or Two) about Safety

A Word (or Two) about Safety

Many people roll their eyes when you mention safety because, of course you are going to be careful! How many times have our mothers reminded us that "it is always fun until someone loses an eye"? And, indeed, fun does screech to a halt once someone has lost an eye. However, safety reminders are warranted in an environment where sparks, coal, and red hot embers of metal will be flying. Although very cool, it's also very dangerous.

So, with that said, you should always wear protective eyewear so that you don't end up losing an eye. In addition to hot flying embers, there will be times when you will use a grinding wheel or other machinery that introduces debris into the air. Protective eyewear should be worn, even if you wear glasses. This will help protect your expensive lenses from getting scratched and pitted from any flying debris.

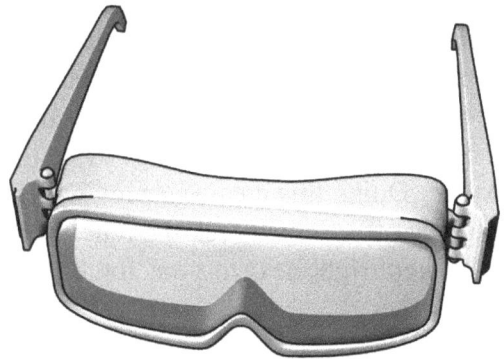

Goggles

A strike of a hammer on an anvil creates a ring that can measure up to 85 decibels, so it is also important to protect your ears. Soft earplugs work well, as do external earmuffs. Keep in mind that your sense of hearing is worth a quick trip to the store to purchase some quality ear protection. Besides having the potential to affect your hearing, noise causes vibrations, which, in turn, causes fatigue. Fatigue can eventually lead to accidents and injuries.

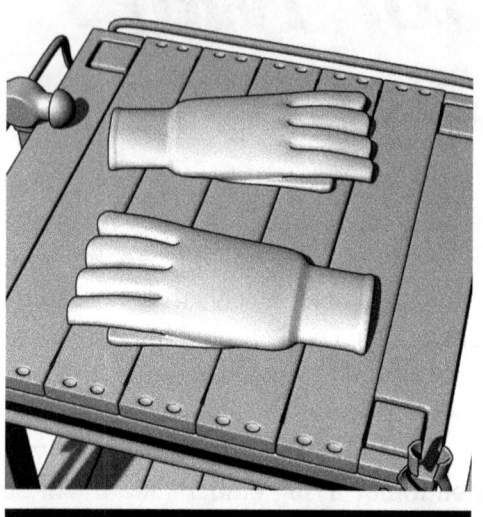

Gloves

Many blacksmiths choose not to wear gloves to protect their hands, which may appear to be a foolish choice for a person who picks up and holds hot metal. Actually, wearing gloves while working with heated metal can be more dangerous for your hands because it is not uncommon for hot embers to fall inside the gloves and trap the hot debris between the hand and glove. If you choose to wear gloves, there are some things to consider when choosing the material.

Cotton gloves tend to catch fire easily, so avoiding them is a no-brainer. Gloves made of synthetic materials (other than Kevlar) should also be avoided as they will melt directly to the skin. Leather gloves seem to work best in the smith shop, but be sure they fit loosely because if your glove is burning it needs to be loose enough for you to flick it from your hand with one quick shake. You must also suppress the urge to submerge your gloved, burning hand into water. Doing this creates steam, which can further burn your skin. Flick the glove off and then seek the relief of the cooling bucket of water that should be kept right next to your forge (we'll talk about this more in a subsequent chapter).

Blacksmiths who choose not to wear gloves should periodically wet down the metal that extends out of the fire to keep it cool enough to grip with an ungloved hand. This can be done using a simple homemade "sprinkler." To make a sprinkler, simply punch about 25 holes in the bottom of an empty coffee can and attach the can to a sturdy stick or metal pole. You will then have a "professional" blacksmith sprinkler.

A Word (or Two) about Safety 21

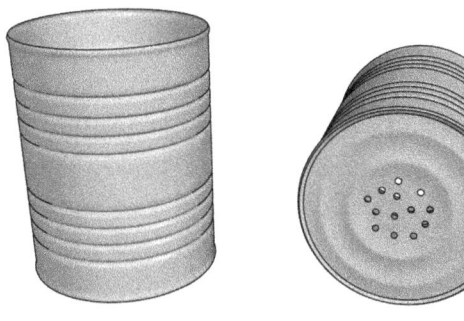

Sprinkler

Note: Whether or not you choose to wear gloves is a matter of personal preference. However, keep in mind that it is not recommended that you wear a glove on the hand that is hammering because a glove on the hammering hand can cause you to loose your grip and send the hammer flying.

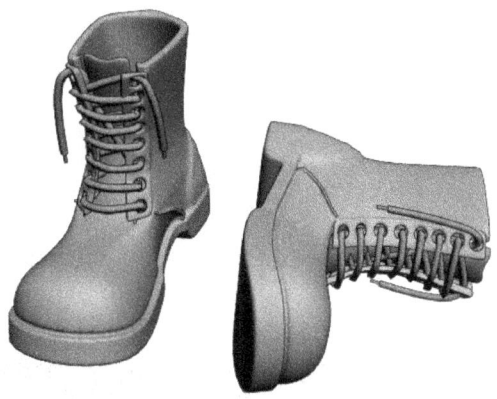

Boots

Footwear is also a concern when working around flying bits of burning material. The same criteria for choosing gloves applies to choosing your shoes. Many athletic-style shoes are made of both cotton and synthetic materials. These shoes can easily catch fire and melt to your foot. Wearing loose shoes is not advisable in the workshop, so you won't be able to kick them from your feet. The optimal choice for blacksmithing footwear would be a pair of leather high topped work boots that prevent burning debris from falling inside the boot.

Contrary to what has been recommended for hand and foot coverings, your clothing **should** be made of cotton or wool. Although cotton burns easily, synthetic materials burn even quicker and can melt and stick to the skin. Sleeves should be short and shirts should be tucked into your pants to avoid being grabbed by rotary equipment. If that were to happen, you would be lucky to only have your clothes ripped from your body. What would most likely happen is that parts of *you* would be twisted into the machinery along with your clothes. Some blacksmiths choose to wear a full-length leather apron. If you find that too cumbersome, a suitable alternative would be to wear a glass cutter's apron, which covers waist-to-knee and consists of a double layer of leather.

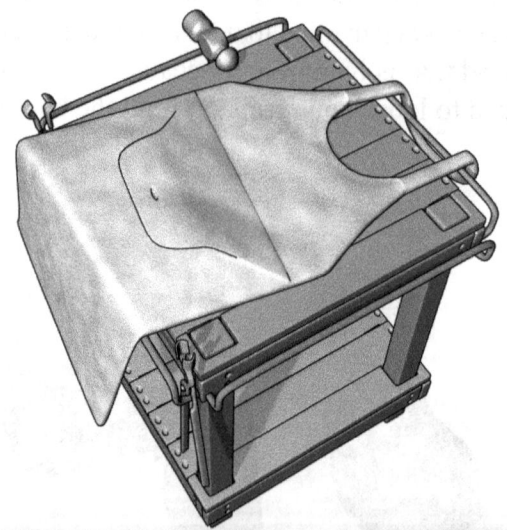

Leather Apron

The grinding wheel is a staple piece of machinery in the blacksmith shop. The stone wheel spins as you shape and finish your new creations or sharpen well-worn tools. If not properly maintained and checked, your wheel can violently explode when you least expect it, sending stone shrapnel in every direction. You can check the wheel for flaws by balancing it in your hands and tapping it lightly with a hammer. What you should hear is a nice "ring." If a dull sound is produced, it is an indication that the wheel may be cracked and should be discarded immediately.

A Word (or Two) about Safety

As an added precaution, allow your grinding wheel to spin for a full minute while you stand off to the side to ensure that the wheel is sound and will not come apart.

To prevent wear and tear on your wheel, always move the metal back and forth while grinding. This prevents grooving. For the same reason, using the side of the wheel should be avoided, or at least only used sparingly and with only light pressure. Your wheel will have a longer life and will be less likely to crack and fly apart if you take care not to mar the surface.

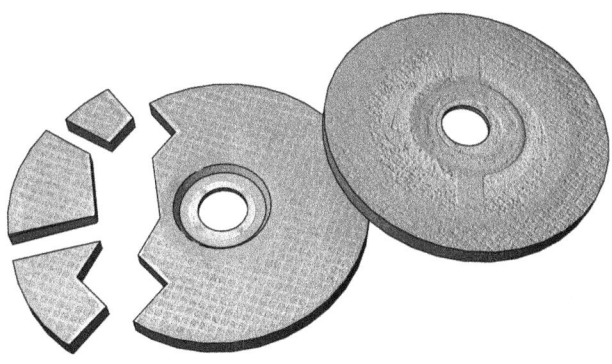

Cracked Grinding Wheel

It is also important to take care to clean the stone between projects. The friction of the spinning wheel can cause some metal residues, when mixed with residue from other metals, to ignite when heated. For example, you may have just ground rough edges from a piece of aluminun and left powder residue on your wheel. Then, the next day, you use your wheel to finish a piece of iron. All of a sudden, you find yourself standing there stunned and smoldering from a very quick, but intense, flash fire. In this scenario, the powder residue from the aluminum mixed with the black powder of the iron (ferrous oxide) became a powerful explosive material when heated. No, you don't need to have a degree in chemistry to blacksmith, but you do need to be aware that such things can, and do, happen. Thoroughly cleaning your wheel will help you to avoid becoming a victim of a flash fire.

Your grinding wheel has a wheel guard. You may be tempted to remove this guard to access more of the wheel's surface, but remember that the guard is there for a reason. The guard will protect your fingers and minimize flying debris should your wheel come apart while spinning. Play it safe – leave the guard in place.

Stretching exercises will help you protect your body from the riggers of blacksmithing and will prepare you both mentally and physically for the task ahead. Blacksmithing provides a great workout, and, as with any exercise, stretching your muscles before you get to work is a great idea. Keeping yourself in good physical condition will ensure years of continued smithing.

Everyone knows that alcohol and drugs have no place in the blacksmith shop. Besides being dangerous, once sobriety creeps back, you'll look at your "creation" and wonder what the heck you were thinking!

A good blacksmith stays focused on his or her work and avoids distractions that can lead to accidents. Keeping your shop a safe working environment is easy when you combine common sense with a good routine.

And, of course, have some fun!

A Word (or Two) about Safety

3 The Shop

The Shop

Setting up your smith shop is as personal as the creations that you will produce there. What works for one person, may not work for another. While many choices will vary according to personal preference, there are some recommendations that we strongly suggest you adhere to.

The floor of your shop can be made of many different materials, but be sure that whatever you choose is nonflammable and comftroable for hours of standing. A flood made of wood or plank should be avoided. Although it would be more comfortable to stand on, you would need to scramble after every piece of hot debris before the floor caught fire. Debris can also fall between the cracks of the planks and smolder for hours before you became aware of the potential fire.

A gravel floor would work well around the forge, but standing on larger stones all day could lead to sore legs and feet. Some smiths like to use a mixture of crushed oyster shells and sawdust. Hot debris is easier to find on this combination because it sends up a "smoke signal", thus making it easy to find.

A concrete floor will not catch fire, but can be very unyielding and cause muscle pain and fatigue through your entire body. One solution is to place a thick rubber mat around the forge for comfort, but be aware that hot debris can burn holes into the mat. A brick floor may serve you well, but brick can also be tiring to stand on for long periods of time. You might also lose bits of metal, screws, and other small materials in the cracks between the bricks.

Natural Flooring

Natural earth flooring seems to work best because it will not catch fire and it is easier to stand on than harder floors. As a blacksmith, you will collect and use many "scrap" pieces of metal. An advantage of natural flooring is that you will be able to rake through it occasionally to retrieve any useful scraps.

Your shop should also have plenty of light. Natural light from lots of windows or skylights would be ideal. Blacksmithing can create copious amounts of soot and dust, and even the best lit shops can quickly grow dark.

The Shop

This leads us nicely into the topic of ventilation. Obviously, where there is fire, there will be smoke! Your shop *must* have adequate ventilation so that you do not become overwhelmed by the soot and smoke produced by your forge. If you plan to smith outside, you'll have lots of ventilation, as long as you remain upwind of the smoke.

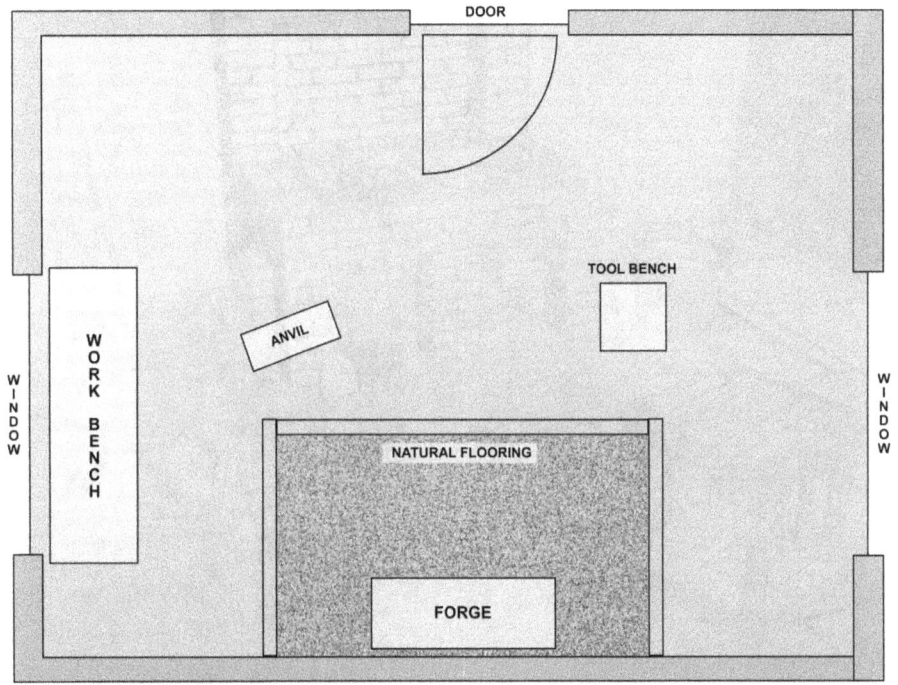

Good Example of Optimal Positioning

Positioning the equipment in your shop is largely a matter of personal preference. You should arrange everything so that you are comfortable moving between the forge and anvil and accessing your work bench and frequently used tools. However, keep in mind that you can rack up a lot of mileage in your shop by taking more steps than are necessary. For example, you shouldn't take more than two steps between your forge and anvil. You will do this often, so you will want them in close proximity. Also, by keeping the motion fluid, you can maintain momentum while crafting. Although we have not yet mentioned all of the components of the shop, the diagram shown is a good example of optimal positioning.

Below is a picture of a blacksmith shop. You can see how closely together the equipment is placed.

Blacksmith Shop

If you take the time to construct a well-organized shop, it will always be standing at-the-ready to serve you. All suggestions offered here are just that – suggestions. You will want to position your shop to fit your own comforts and needs. The only exception would be the recommendation to avoid a wood floor. You don't want to find yourself scratching your head and wondering what happened as you watch the firefighters extinguish the shop you worked so hard to build.

The Shop

4

Equipment

Equipment

By now you have decided where you will be smithing and your shop is ready for some equipment. The recommended equipment for the beginner's workshop is rather basic and you may not need everything that is presented here. It will really depend upon the complexity of the creations you will craft.

A well-equipped shop should include the following: a forge, an anvil, a sturdy workbench where various equipment can be mounted, a tool table, a quenching bucket, and a coal bin. Although you are surely going to want to make your own smithing tools, you will obviously need to start with some store-bought ones before you can craft your own.

There are plenty of hardware stores and specialty shops where you can purchase your smithing equipment, but this can be very expensive if you are starting from scratch. It is perfectly fine to equip your shop with second-hand articles, as long as they are in good working order. Flea markets and garage sales can produce a wealth of surprising finds that will serve your shop well. You are also lucky to be living in a time when the Internet is at your disposal. There are many "blacksmith associations" online. These organizations are a great place to "meet" other blacksmiths, and many of the websites contain areas where used equipment is posted for sale or trade.

The Forge and Fire

The forge is the heart of your shop because it contains the fire that will allow you to bend steel. Fire lives and breathes and is fed by oxygen, which *you* will have to control. *You* will decide how hot the fire will get and when to cut off the air when you think it has had enough. Rather empowering, isn't it? Now let's build that forge!

As previously mentioned, you could probably find a used forge, or build your own, but where should you place it? Forges can be either round or square. A square forge will allow you more space in your shop because it can be pushed against a wall. Whichever you choose, you will *not* want to place the forge in the center of your shop because this would limit your work space and potentially impede your movements. A centered forge may also allow air to blow in all directions. This could cause your fire grow unexpectedly, or blow hot cinders into your face (yet another reason to wear safety goggles).

You will want to place your forge in the darkest part of the shop because you will need to be able to see the color of the metal as it is heated. If you are forging in broad daylight overheated steel will be difficult to see, and if not controlled, could damage the integrity of the metal. An outdoor forge should have a hood to create shade from the sun.

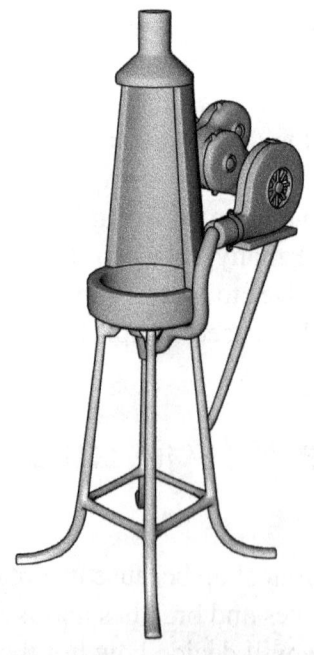

Hood to Shade against Sunlight

Equipment The Forge and Fire

The basin of the forge needs to be large enough to hold a sufficient amount of glowing coal. This is where you will heat the metal. The basin can be made from many different materials, such as brick, stone, or fireclay. Most modern forges have a basin made of steel or cast iron. A comfortable height for the forge is determined by the height of the smith. Typically, the edge of the open hearth is 30 inches from the ground, but adjustments should be made for your own personal comfort.

There must also be a means of introducing air into the fire, which can then be controlled to allow for more or less heat. In the Middle Ages, bellows were invented and served as a great tool to fuel a fire. They worked so well that this method is still used today. However, controlling bellows while working the fire takes some practice. It can also be difficult to maintain a steady flow of air when needed because the bellows must refill with air as it is pumped. If you had a helper (or an extra set of hands), two bellows could be pumped in turn, so that the air flow remains constant.

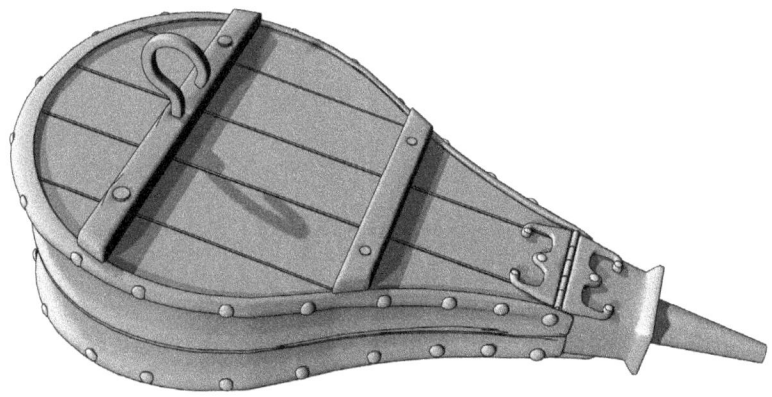

Bellows

The hand-cranked blower is a more popular method used to create airflow. Your free hand will still be occupied, but you will be able to create a constant flow of air. As an added benefit, when you stop cranking, the air stops immediately. This method can help you to be the master of your fire.

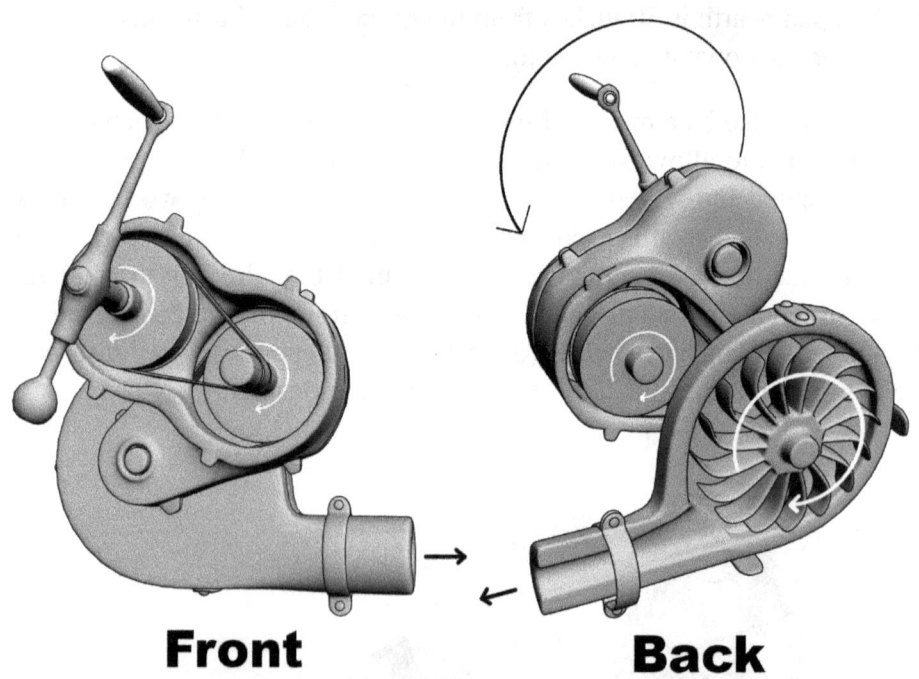

Front   **Back**

Hand-Cranked Blower

As technology advances, so do the choices for today's blacksmith. Electrically driven fans are quite popular with the modern smith as they can continually feed the fire without interruption. The electric fan can also be controlled with a foot pedal, thereby freeing up both hands. It is advised, however, that a rheostat switch be installed. This will allow you to stop the air flow as soon as your foot is removed from the pedal. Using this system without the safety switch will require you to remember to shut off the air flow each time you begin forging your heated metal. If you forget just one time, you risk having an ever growing fire at your back.

Equipment The Forge and Fire 41

In addition to the fire hazard, if you have to stop and manually turn off the air you will lose precious hammering time when the metal is at the desired temperature.

Ideally, the air should be blown from under the fire. This allows for even air coverage and will help to sufficiently heat the fire. The first example below shows a simple inlet pipe with a couple of perforations (these are easy to create). The second example depicts a commercial air system.

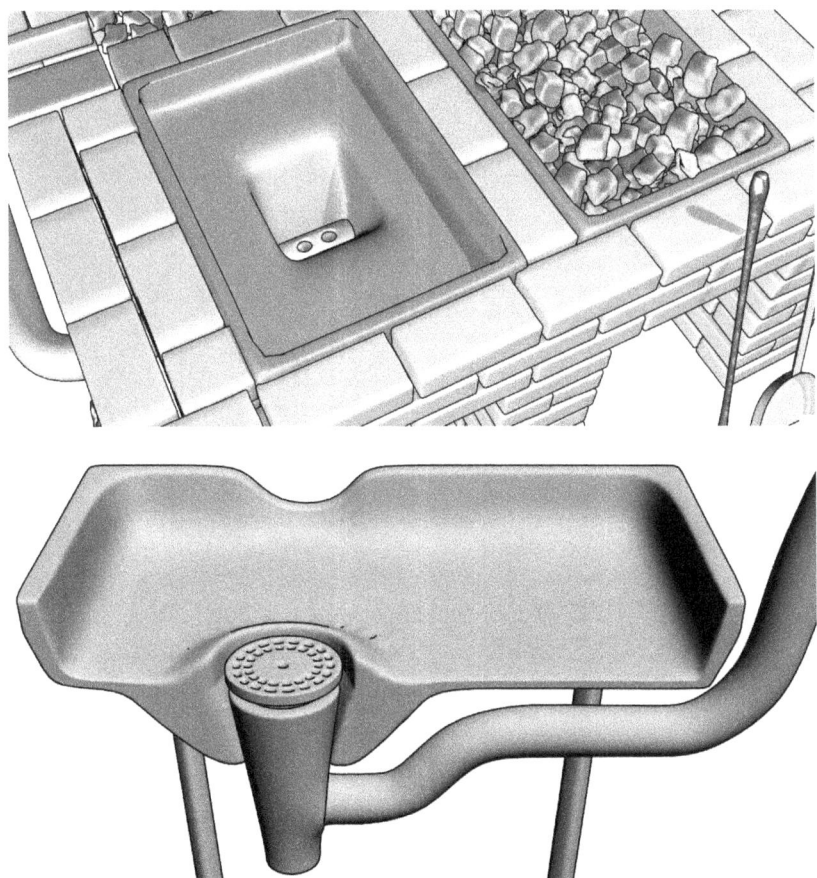

Air Systems

Most forges operate under the same basic principles. The picture below demonstrates how a flue takes smoke away from the fire. A proper fire hood contains a "smoke shelf" with an opening about ¾ the size of the hood. This shelf essentially acts like a dam on a river. Just like a dam causes a river to flow faster downstream, the smoke shelf causes hot air to flow quickly and smoothly past the shelf and out the exhaust. This strong draft of air pulls the smoke up and away from the fire. This is affected by Bernoulli's Principle. Physicist Daniel Bernoulli found that as the speed of moving fluid increased, the pressure within the fluid decreased. The same holds true for air under the same circumstances.

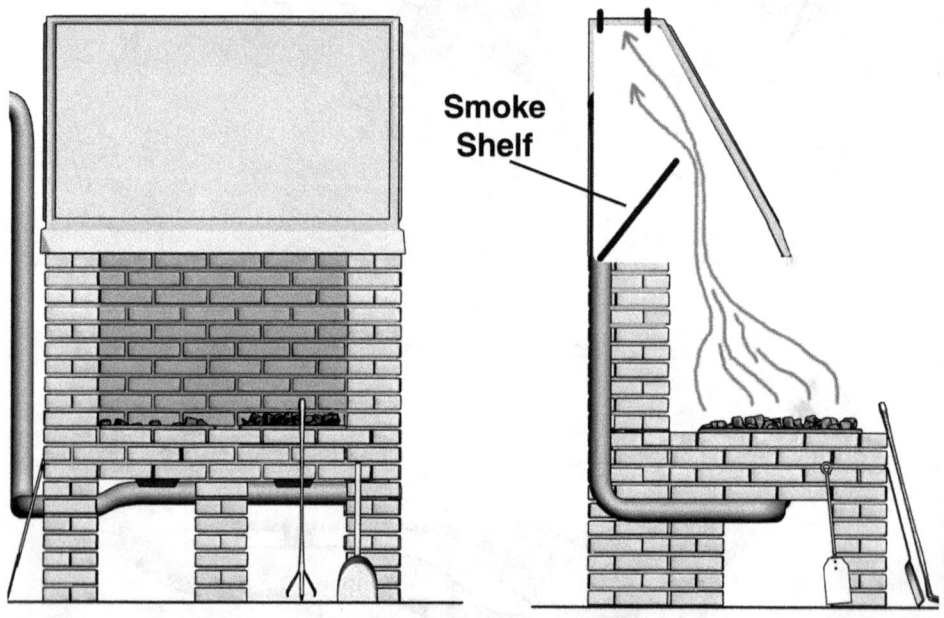

Flue Takes Smoke away from the Fire

Once again, advances in technology give the smith more options. A hinging smoke-catcher is a modern addition to the forge and works like the anti-pollution devices found in some automobiles.

Equipment The Forge and Fire

In cars, the device will catch the smoke created from the oil and funnel it back into the combustion chambers of the engine. For the forge, the device guides the smoke back into the fire, via the air intake. Once the fire starts, the flames consume the smoke itself. The "smoke catcher" can be swung on its hinges and out of your way. The existing fire will keep itself relatively smoke-free.

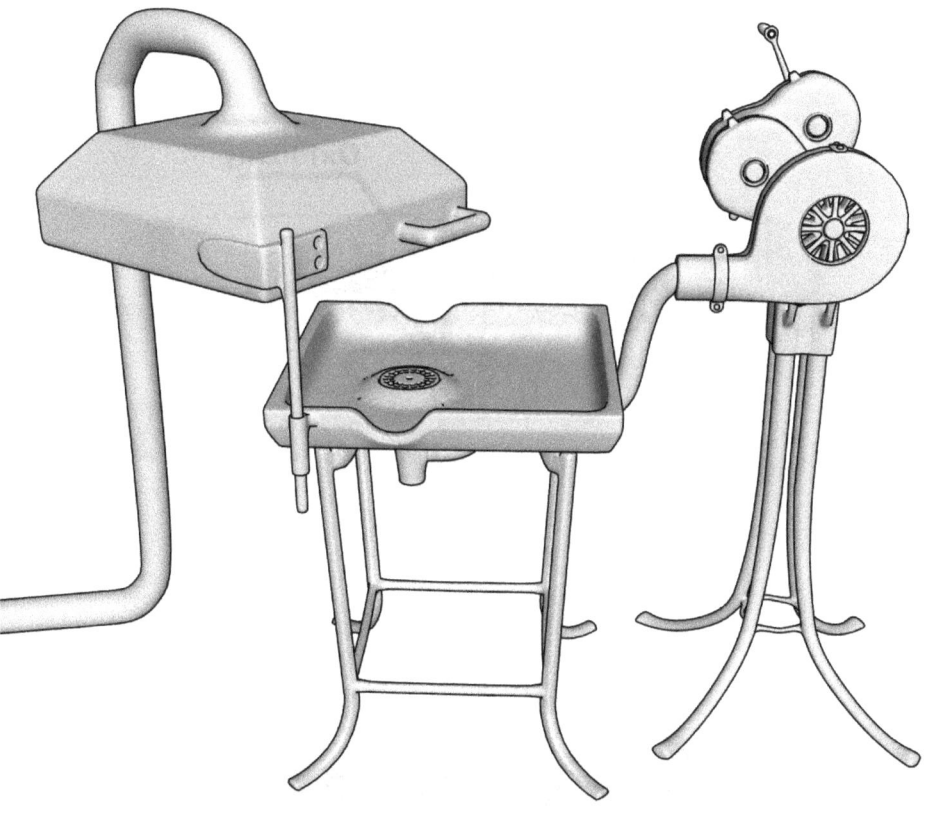

Smoke Catcher

Proper understanding of how the forge works will help you to maintain a clean and medium-sized fire. Maintaining the fire so that it remains effective all day is probably one of the more difficult skills the beginner smith will have to master. But, practice makes perfect. In no time at all you will find the process routine and will be managing your fire like an old pro.

The most common error made by new smiths is that they tend to build and maintain fires that are too small. To be effective, the fire should be relatively large and deep. This allows different and distinct levels to form within the fire. These levels are known as the Oxidizing, Neutral and Carbonizing levels. The pictures below illustrate where the different levels form within the fire. You will heat your metal in the Neutral level.

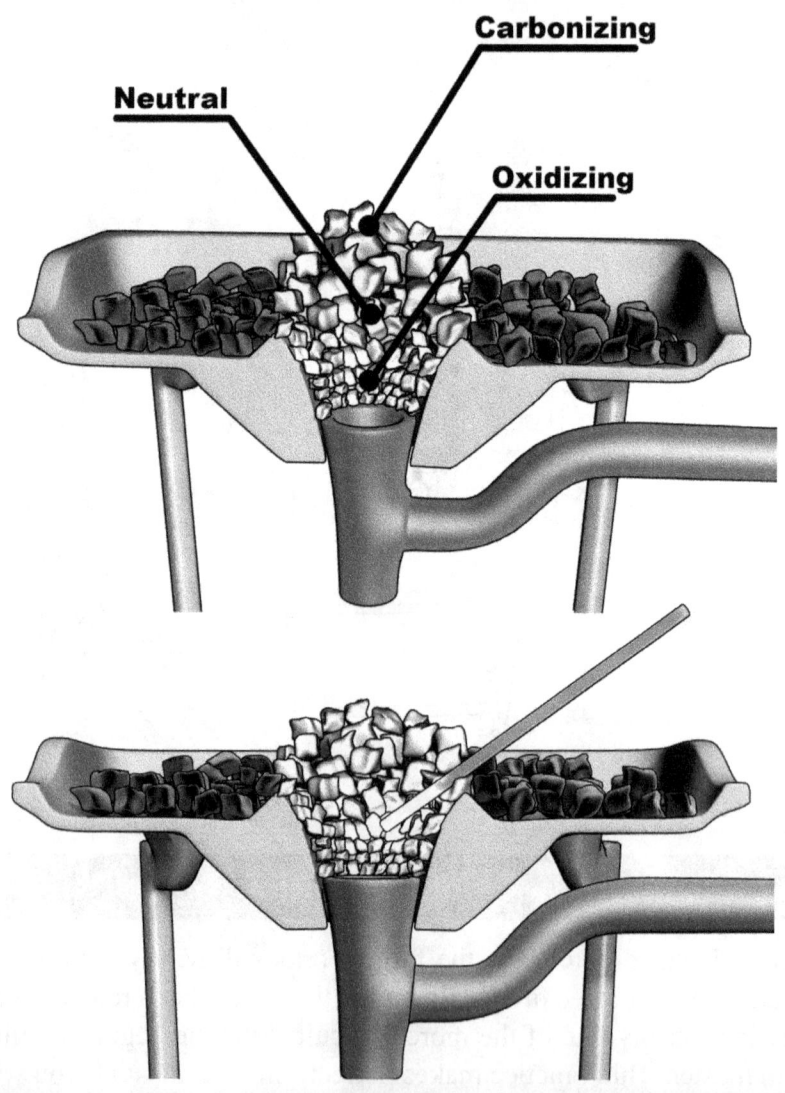

Levels Forming within the Fire

Equipment The Forge and Fire

The most typical fuel for the smith's fire is blacksmith coal (sold commercially as metallurgical grade coal). No, this is not the same as the charcoal briquettes you use in your barbeque grill. It is a much smaller, pea-sized coal. As previously mentioned, the blacksmith's fire should be kept "clean." If the flames of your fire burn yellow, your fire is "dirty." The yellow flames indicate that something is burning in your fire that is not pure carbon. This can damage the integrity of the metal you are heating. The "extra stuff" burning in your fire is likely remnants from past fires and projects. For example, improperly heated metal will oxidize and bits of burnt metal will flake off and fall into the fire. These flakes sully the current and subsequent fires. Avoiding the flakes by properly heating the metal will be discussed in future chapters.

You might imagine that where there is a large fire, there must also be a lot of smoke. This rule does not hold true in the smith shop. A properly built blacksmith fire will be relatively smoke-free. If your fire begins to smoke, it is most likely because the fine ash debris from your smithing has fallen to the bottom of your fire and is blocking the air delivery system. If you haven't cleaned it out in a while, the fire may smoke because it is being choked by all the ash. Time to clean out your forge! If you purchased a commercial forge, it will most likely have an "ash door" that can be opened to allow the ash to fall from the forge. If you build your own forge, you should strongly consider equipping it with this feature.

Coke is a by-product from the coal as it burns in the fire. As the burning coke reduces to ash, you must feed the fresh forming coke back into the fire from all sides. You will then add fresh coal to take its place. This procedure of feeding coke to the fire and adding coal will generate a fire that is clean, hot, and nearly smoke free.

To start your fire, you can use kindling, such as wood chips, or you can use the coke left over from the previous fire. Pile the kindling of choice into the center of the forge basin, leaving a small space in the middle. You will then fill the surrounding area with your blacksmith coal.

The coal must be *wet*. This will help the coal to pack more tightly and allow the fire to burn with more intensity. A ready supply of coal should be kept on hand to be raked into the fire as it grows and to replace what the fire consumes. This supply of coal must be wett down occasionally to keep the fire from spreading unintentionally. This is where your "professional" blacksmith sprinkler, mentioned in a previous chapter, will come in handy. Just sprinkle the coals from time to time. Once the fire's stage is set, crumple a piece of paper into a tight ball. Light the ball of paper and hold it in the center of your kindling (with a poker, duh). Next, start your airflow very gently, feeding the infant fire with enough air to encourage it to grow, but be careful not to let it extinguish. As the fire grows, increase the air. When the paper ball is burning well, begin to rake your kindling over it. Just the kindling at this point. Continue blowing air into the fire with increasing force as the fire grows. There will probably be a considerable amount of smoke during this process, so don't assume you are building a "dirty" fire.

When the kindling becomes evenly hot, you can begin to rake in small amounts of fresh coal. Allow your fire to grow – don't try to force it. *Do not poke and stir your fire.* It's not a pot of stew! You want the coals to remain tightly packed. Pushing them around will make them try to work individually, rather than as a team. It's a bad habit that is best not to start.

The Anvil

The anvil is the piece of equipment that will take the most abuse. Many early anvils were made of iron, but they did not stand up well to relentless hammering over time. Faithful service over many years left them looking misshapen and sad. Most modern anvils are made of steel and are well suited for their intended task.

Although most blacksmiths like to make their own tools, the anvil is one tool that will need to be purchased from a specialized manufacturer. There really is no substitute for a proper anvil.

Equipment The Anvil

Many traditional blacksmiths prefer to use an anvil that sounds like a bell when struck. While there is no real proof that a beautiful sounding anvil works better than one that sounds off-key, one that can carry a tune is nicer to be around for long periods of time. You are sure to find an anvil that strikes a chord with you.

The anvil's shape has changed over the years and some are equipped with more features than others. If an anvil is used to craft more delicate creations, its appearance will be very basic and may have customized edges for specific shaping. These smaller anvils typically weigh about 50 pounds. In contrast, a commercial smith shop that performs work on heavy equipment may use an anvil that weighs 800 pounds. For general smithing, the anvil will most likely weigh between 100 and 200 pounds.

Most modern anvils, including the one pictured below, conform to a shape referred to as the Londen pattern.

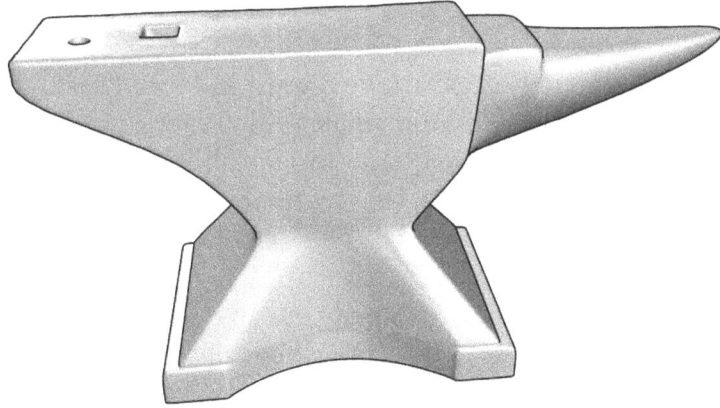

Anvil

The main working surface of the anvil is called the *face*. The face steps down to an area referred to as the *table*. If you decide to buy a used anvil, check to be sure that the edges of the face and table are sharp and straight. The edges may have been deliberately rounded for specialty work, but typically they should be well defined. The table is the area used for cutting steel, and its surface is purposely more yielding so that it does not damage the cutting edges of your tools. Be sure to examine the table of a used anvil to ensure that it is not overly compromised by saw cuts. The surface can be corrected by filing and grinding, but once metal has been taken away, it can not easily be put back (it is possible to weld metal back onto the anvil, but it is a tedious and difficult task).

From one end of the anvil juts the *beak*, or *horn*, which can be used to round pieces of metal. The opposing end is the *heel*. The heel will usually have two holes running straight through the metal. One hole is typically square and is called the *hardie hole*. It derives its name from the hardie tool because it fits nicely into the hole. The other hole is referred to as the *punch* or *pritchel hole* because it provides clearance when a punch is driven through hot steel. The heel may actually have several pritchel holes for you to choose from.

The body of the anvil is supported by a waist and broad base. The base is designed to prevent the anvil from tipping. However, a safe workshop will have a securely anchored anvil. The form of the feet provide the perfect shape for securing the anvil with bent spikes, and some may even have pre-punched holes for this purpose.

You must now choose a base to mount your anvil upon. A wooden base is fine, as long as it will sufficiently support your anvil. If you plan on using a wooden base, be sure to use thick, bulky wood that will withstand the weight of the anvil and the powerful hammering. This type of base is especially good if you smith in different locations because it is fairly portable.

Base for the Anvil

There are also pre-fabricated stands made of cast iron. These stands have a recessed surface for the anvil base and are similar in shape to the wooden bases. Cast iron bases have never really gained much popularity because the metal anvil atop the metal base increases the noise level of the shop substantially. Aside from being noisy, the cast iron base is unyielding and may cause the anvil to rebound under heavy hammering. The anvil and base should never move when you hammer. An added benefit of wooden bases is that leather straps can be attached to the wood and tools can be hung from these straps.

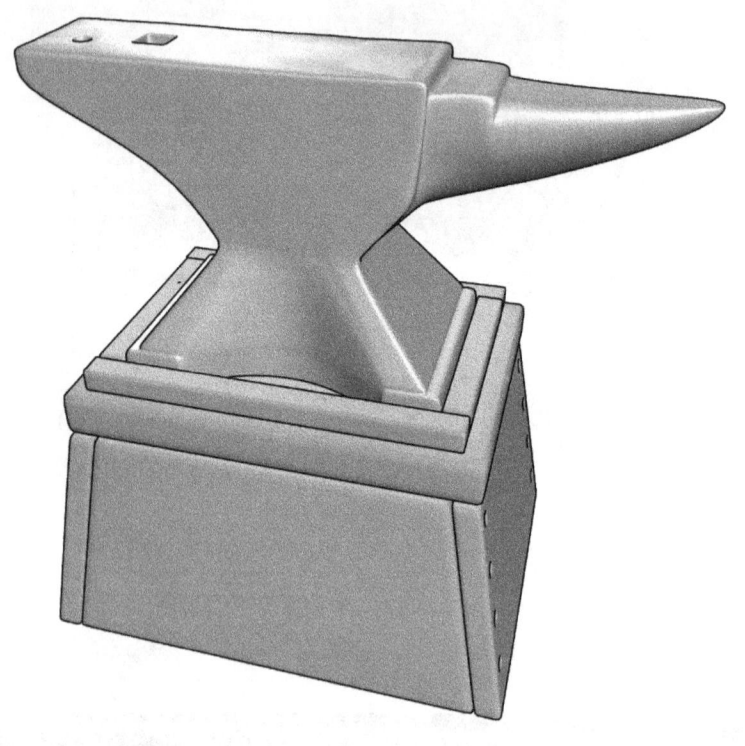

Anvil and Base

For the permanent smith shop, a section of tree trunk is the anvil base of choice for most blacksmiths. The block should be ample enough to accommodate the entire base and the feet of the anvil. The tree trunk will need to be buried at least 3 feet into the natural flooring of the shop to prevent the whole thing from toppling when you hammer. If you decide not to take this precaution, make sure that your leather work boots have steel tips so that your feet are not crushed when the tree and anvil land on them.

Equipment The Anvil

Becuase people are different sizes, the following is a useful trick to determine what anvil height is best for you. By using this guide, the anvil placement will also be the correct height to allow you to come down with a forceful blow of your hammer when you need to do so. A comfortable height will allow you to stand almost upright when holding a piece of metal horizontally across the anvil. You don't want to be bent over, sacrificing your back for your craft. To judge the correct height, stand next to the anvil and base with your arms at your side. When you make a fist, your knuckles should rest comfortably on the face of the anvil. This will determine the perfect anvil height for you.

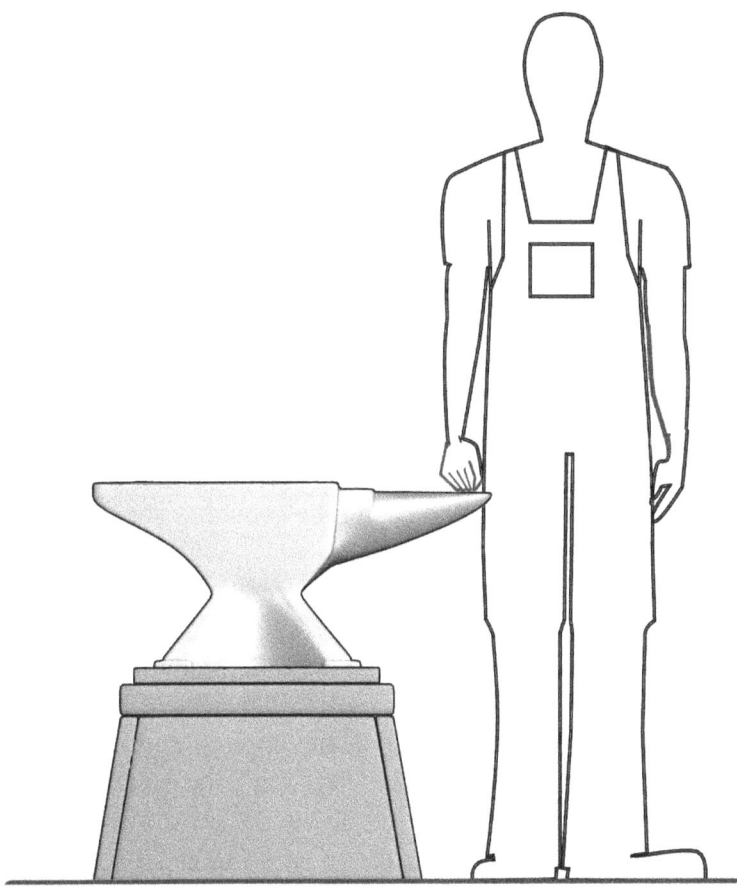

Correct Height of Anvil on Base

Be sure to appreciate all of the hard work and abuse that your anvil endures. It may seem a bit silly, but many blacksmiths tap their anvils with a hammer three times as their last task of the day. So, as a sign of your appreciation, give the face or your anvil a few good taps as you leave your shop.

The Quenching Bath

The quenching bath should be kept right next to your forge and will be used for several purposes. The quenching bath is typically water, although some smiths prefer other liquids (we'll mention those in a moment). You can use a simple water bucket for your bath, but a large rectangular metal container will work best for dunking larger pieces of metal. The bath should not be smaller than 10 x 24 inches and no less than 15 to 20 inches deep.

The purpose of the quenching bath is to quickly cool hot metal. Your professional blacksmith sprinkler should also be hung either next to the bath, or right on its edge. The sprinkler is used to keep the coals around the perimeter of your fire wet. This ensures that your fire does not unintentionally grow. It will also be used to keep any metal extending out of the fire cool enough to handle. If hot metal happens to fall onto wood, sawdust, or fabric, the sprinkler will be ready to prevent a catastrophe. Let's not forget that the quenching bath is also essential in the event that *you* happen to get burned! Burning skin should be cooled as soon as possible to prevent further injury.

Some smiths believe that they get the best result from quenching metal in brine. Brine is made by saturating water with common rock salt. Oil or lamb's fat is great for hardening very small delicate tools. The lamb's fat will leave quenched metal remarkably clean, whereas motor oil will leave some black carbon deposits on the metal. Although it requires a bit of work to remove the deposits, the metal is left unharmed by the process. If you plan to use this type of quenching bath, be sure to keep a tight lid on the container to keep any rodents away.

Whichever bath you choose, you will still need to have water on hand for purposes other than tempering the metal because it is not a good idea to toss motor oil or lamb's fat on an uncontrolled fire . . . .

The forge, anvil, and quenching bath should all be kept in the darkest corner of your shop. This is necessary because you will need to be able to see the color of the metal you are heating, hammering and quenching. Overheating or under heating can change the integrity of your metal, causing your creation to turn out in an unexpected way.

The Coal Box

You will also need a container for your coal. It should be kept close to the forge, but not so close that it could unintentionally catch fire. The ideal location is under your quenching bath. If your bath container is flush with the ground, then the coal box can be placed on the opposite side of the bath, away from the forge. Remember to periodically wet the coal in the box to keep it ready to go into the fire.

The Tool Table and Workbench

On the right side of the anvil (or the left if that is your dominant hand) you should have a small tool table that measures approximately 20 x 20 inches. Ideally, your tool table will have an iron bar around the sides to give you an extra place to hang tools. The table could also have a bottom shelf for storing large or oddly shaped tools. The following picture is an excellent example of a "user friendly" tool table.

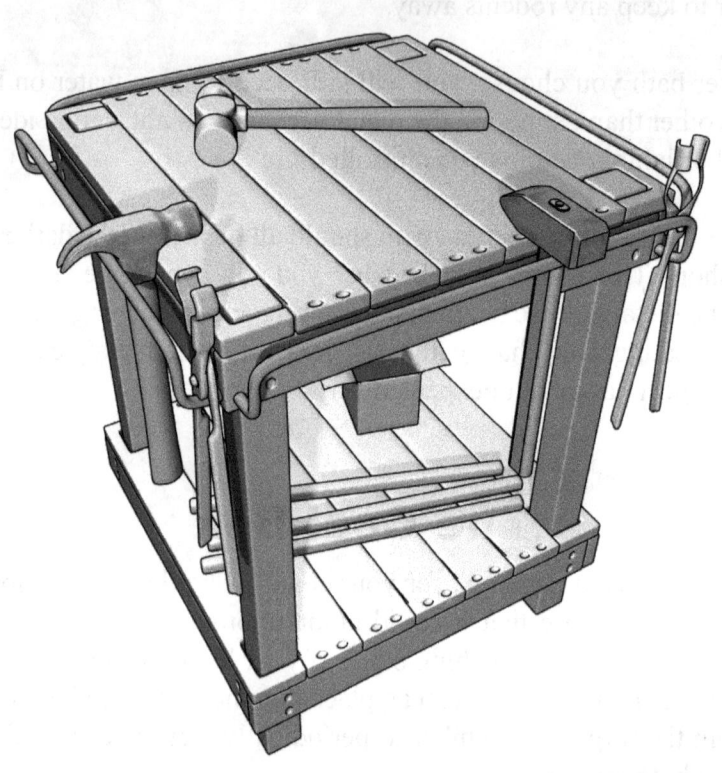

Tool Table

Think of your workbench as you would your anvil. It will need to be sturdy, unwavering, and secure. You will have heavy equipment mounted to your workbench and will often continue to hammer your creations on it. The bench needs to be capable of enduring a lot of abuse. A long, roomy bench is ideal; but the suggestions offered here are for those smiths who have limited space in their shop. Be sure that all equipment is securely mounted and will not wobble when you hammer.

Equipment The Tool Table and Workbench 55

Workbench

The Drill Press

The drill press will be one of the most important pieces of equipment in your shop. For heavy metal working, you will want a sturdy and decently sized press. When bolted to the bench, the swivel table should be adjustable in every position and high enough to safely clear any other mounted equipment when you are working on a long piece of metal.

One of the most versatile machines in the smith shop, the drill press can do routing, filing, rasping, grinding, and, of course, drilling. It can even be adapted to function as a wood lathe. Once you have mastered using your drill press, your smithing will be bound only by your own imagination!

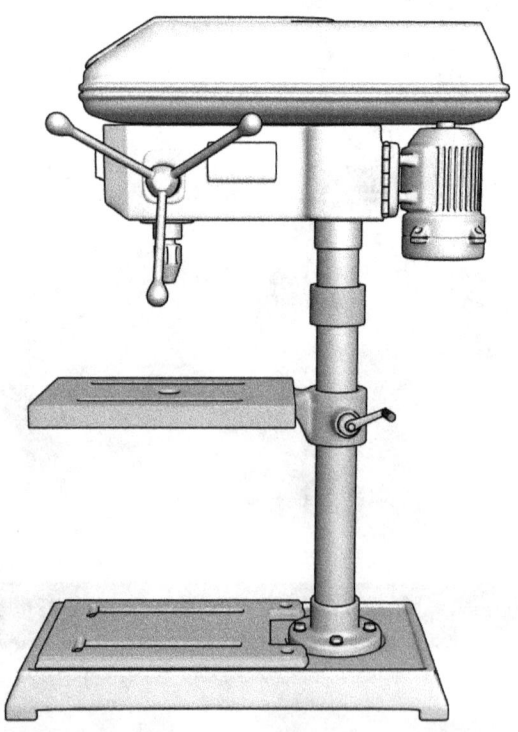

Drill Press

No drill press is complete without a drill press vice. The vice is placed on the swivel table of the press and is designed to hold metal in place. The metal should have already been marked by a center punch. The vice will hold the metal firm, and the punch marks will guide the drill to ensure it does not wander off its intended mark.

The Grinding Wheel

No smith shop is complete without a grinding wheel. The grinding wheel is essential to finish shaping your beautiful creations. A slower grinding wheel is recommended for blacksmithing purposes. One that provides a speed of 1750 rpm may safely spin an 8 inch diameter, ¾ inch thick wheel. For higher speeds, a smaller wheel needs to be used. If your metal contacts a high-speed wheel off-center, the instrument can kick, or fly apart. The grinding wheel is definitly a tool to use carefully as there is also the added risk of flying sparks or debris hitting your eyes.

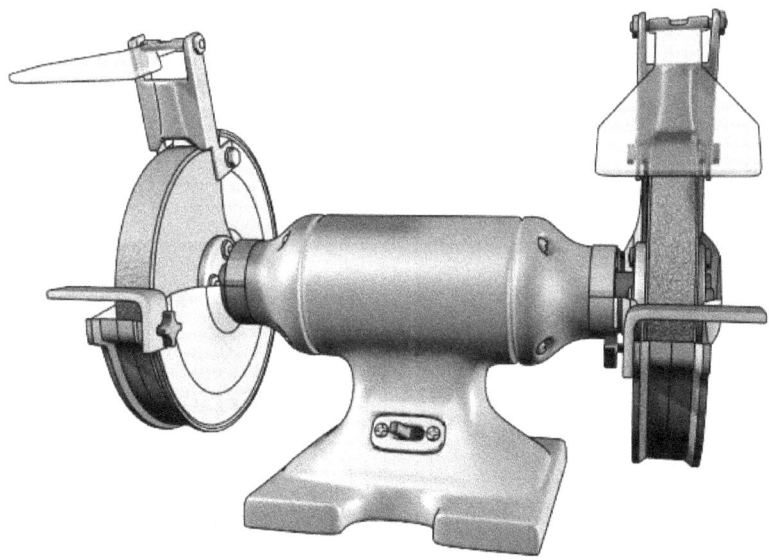

Grinding Wheel

The Abrasive Cut-off Wheel

If you have a circular table saw, you can substitute your saw blade for an abrasive cut-off wheel. This will enable you to cut hardened steel. The wheel will cut the metal cleanly and easily. This is not an essential piece of equipment for your shop, but it can sure come in handy. An abrasive cut off wheel is especially useful if you are a beginner blacksmith whose main source of material is scrap metal.

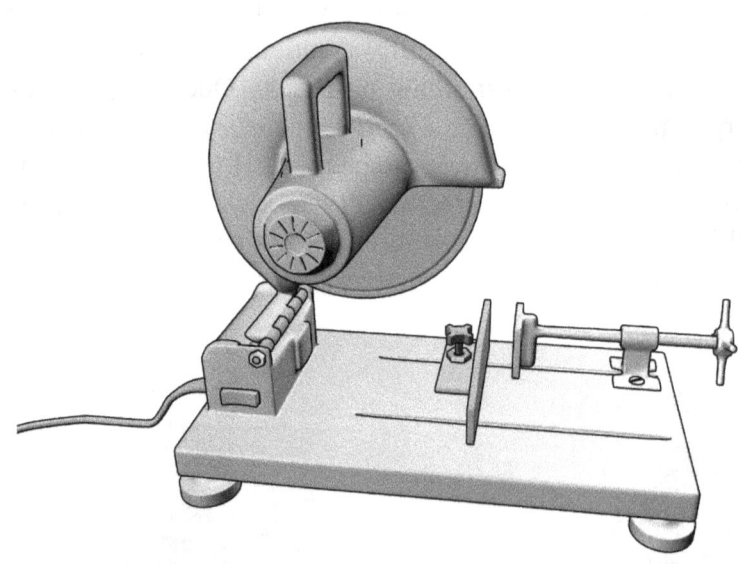

The Abrasive Cut-off Wheel

The Cotton Buffer

The cotton buffer is also not an essential piece of shop equipment, but it merits mention and you may find you can't live without it. You can buy a ready-made buffer attached to a small shaft as a chuck insert. It can be clamped onto your drill press or grinding unit, and is used to finish and polish your metal creations. The buffer, when used in conjunction with a buffing compound, will finish your creations to a high shine.

Equipment The Cotton Buffer   59

The best buffing compound is an abrasive wax called *tripoli*. It should be available in almost any hardware store. When the compound is held to the spinning cotton wheel, the friction will heat the compound and the wax will melt onto the buffer. Once you allow the wax to cool and harden, the buffer will polish your creation to a mirror shine. Have your sunglasses nearby!

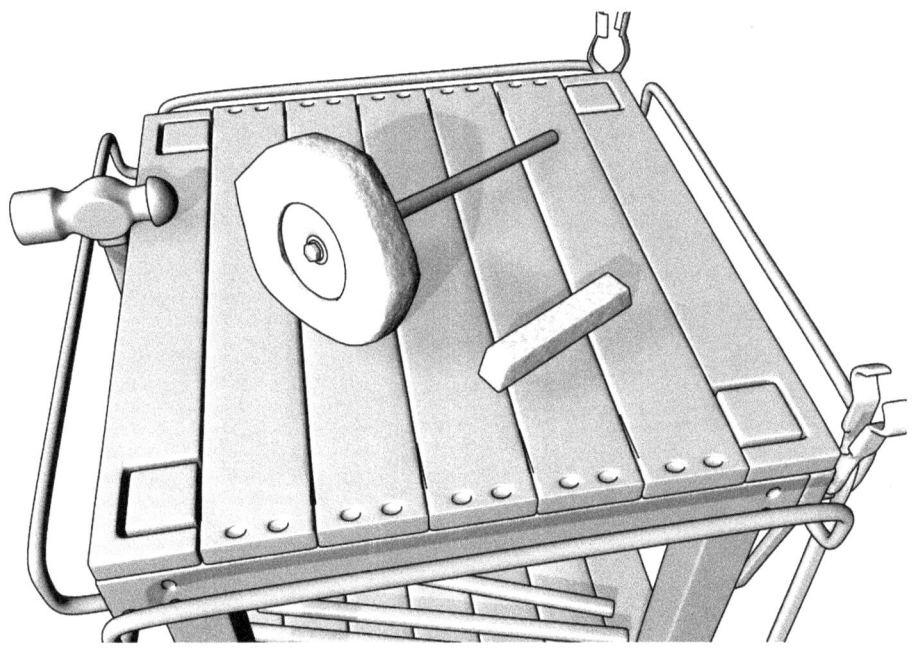

The Cotton Buffer and Buffing Compound

As mentioned at the beginning of this chapter, you may not need all of the equipment mentioned here, or you could find that you need more. All smith shops require some very basic equipment, but your shop will be built to suit your personal needs. Don't be afraid to experiment with the scrap metal you have collected. The beauty of blacksmithing is that if you mess something up, you can just heat it up and start over!

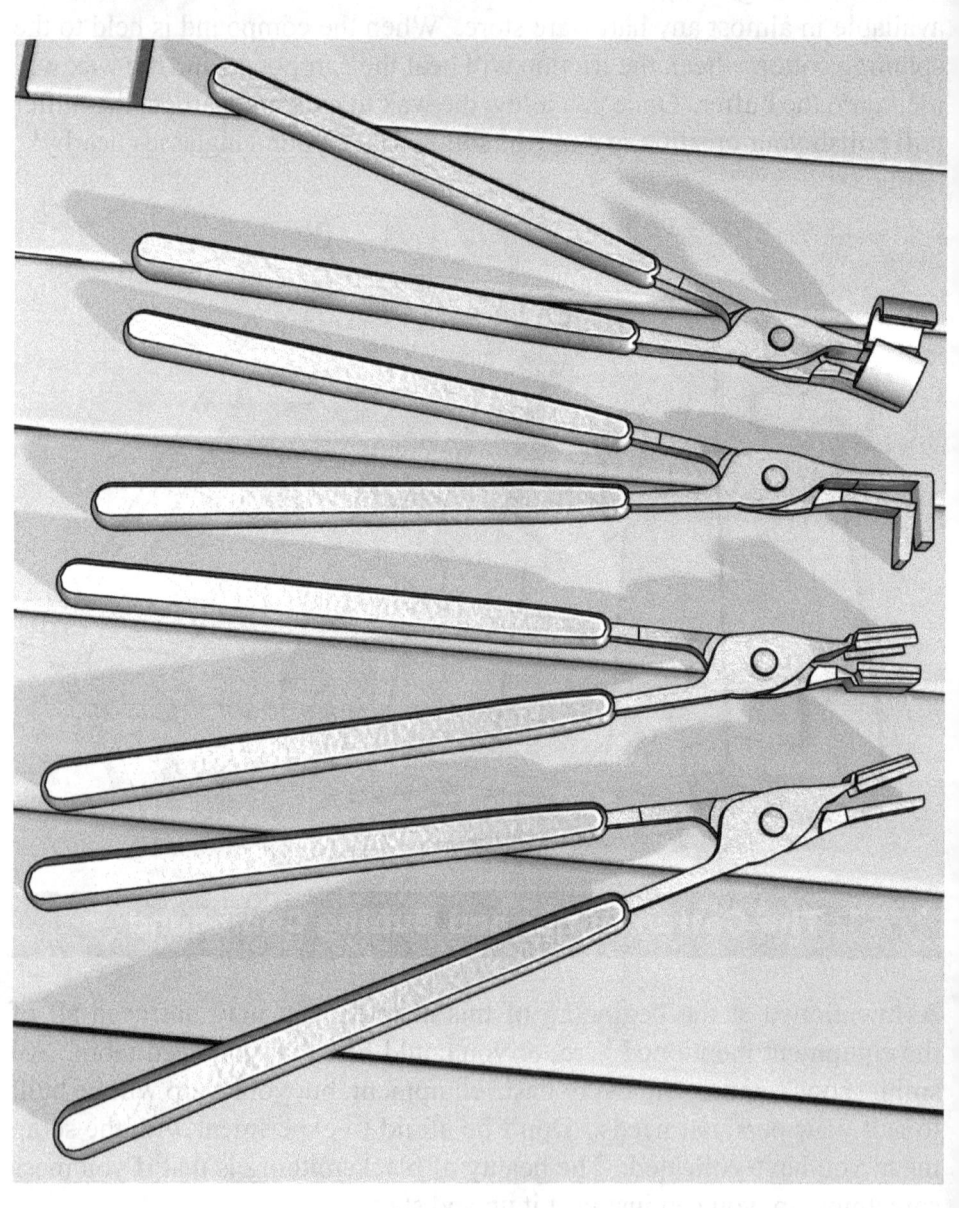

5

Essential First Tools

Essential First Tools

The previous chapter described several tools commonly used by the beginner blacksmith. While all of these tools will help you to realize the vision of your creations, every blacksmith needs a good hammer. Most of the work you do will require hammering, whether it be striking the hot metal directly, or striking a tool held over it. Sure, the "fancy" equipment can be invaluable when you need to perform intricate work, but the base of all your creations will be formed by your own muscle power.

It can be confusing to know what types of tools you should purchase for your shop, as there is a tool for just about every purpose. By presenting various options here, we hope that you will be comfortable with your initial choices. As you become more experienced, you will form your own ideas about the types of tools you will need.

Hammers

No smith shop is complete without at least one hammer, and you will likely accumulate many. Of course, you will eventually want to make your own tools, but your first tools will need to be purchased. After all, you can't make tools without tools! As in gathering the other equipment for your shop, your first tools can be either store bought or second hand. As a general note, the handles of your hammers should be made of wood. Some hammers are constructed so that the head and shaft are formed from a single piece of steel. Constant hammering with this type of construction will transfer the shock and vibrations from the head of the hammer to your hand (and the rest of your body). Remember what was said in the safety chapter about vibrations and fatigue?

You will find that there are a variety of hammers for you to choose from, depending on what you are crafting. Different hammers are made to accomplish different tasks. A good general first choice is an engineer's ball peen hammer, which is pictured below. This hammer sports a flat (or slightly domed) end, and the opposing end is nicely rounded. You will find more uses for the flat side of this particular hammer, although the rounded side is ideal for riveting. Hammers that you will find occasional uses for are the straight peen and cross peen, which are both pictured below.

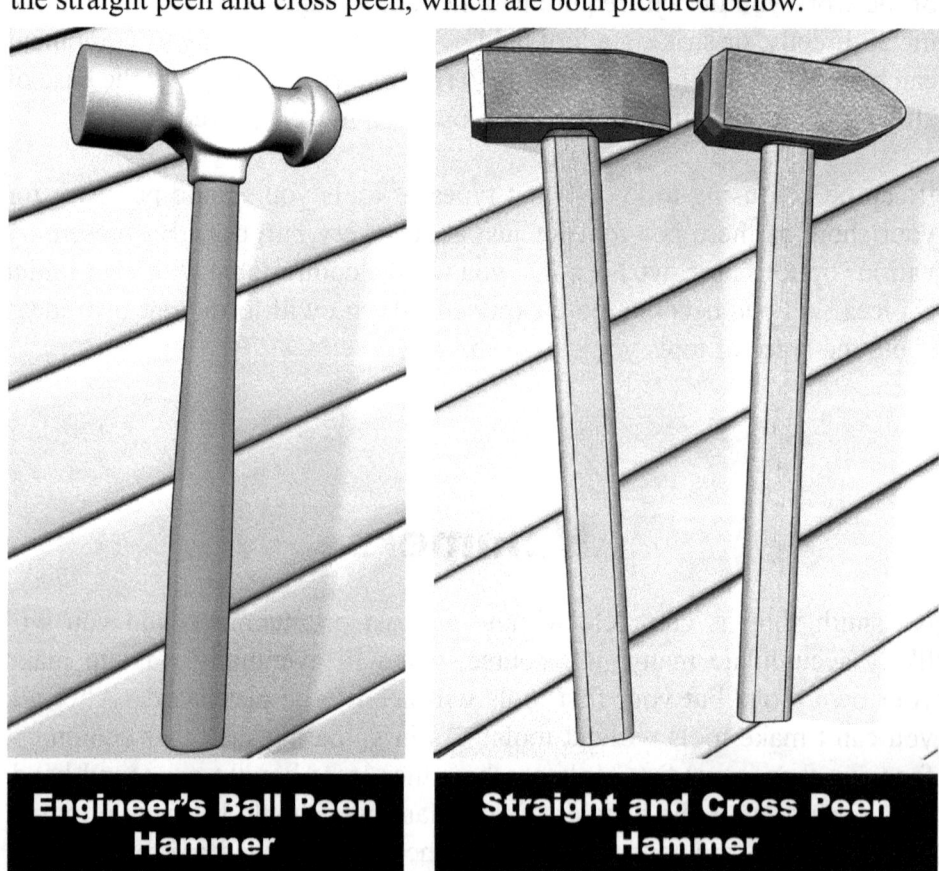

Engineer's Ball Peen Hammer

Straight and Cross Peen Hammer

The hammers pictured above are ones that the blacksmith will typically make on their own. Straight and cross peen hammers are most often used to stretch or draw out heated metal. Although they complete the task effectively, you will most likely finish smoothing the metal with a flat-headed hammer. An example of drawing out metal might include flattening heated steel into a shovel, or widening and tapering the end of a heavy knife.

Essential First Tools Hammers

You will notice that the edges of the straight and cross peen hammers are turned in different directions. This design allows you to strike the metal comfortably, while avoiding the need to turn your project to accommodate the angle of the hammer's edge. Are you familiar with the saying "Strike while the iron is hot"? It is important to shape your creation while the metal is at the correct temperature. Having both hammers close at hand will allow you to do just that. You will also appreciate not having to contort your body to align yourself with the edge you are flattening.

A sledge hammer is also useful in the smith shop, but is typically only used if you have a helper. The weight of a sledge hammer can range from 4 pounds to 14 pounds, so you can imagine how difficult (or impossible) it can be to handle your creation with one hand while swinging a sledge hammer with the other. If you *can* swing a 14 pound sledge with one hand, you probably won't encounter too many people who would criticize your technique! However, for most folks, a 7 pound sledge is a good starter weight.

Hammer sizes, no matter their style, are usually denoted by their weight, which is typically up to three pounds. A good starting weight is a 2 pound hammer, but that can vary for the individual smith. Any hammer weight can perform well as long as it can be controlled by the smith. Most importantly, choose a weight that is comfortable for you.

Hardies and Sets

There will more than likely be a time when you will be working on a project and will need to cut a piece of steel. Although a hacksaw is sometimes used for the job, most quick cuts can be accomplished with a mighty blow of your hammer. A strong blow should either cut the metal clean through, or enough of the way through to allow you to snap the piece in half. To assist you with this task, you will want to keep a supply of useful tool "adjuncts" in your shop.

If you are going to walk the walk of a blacksmith, you should also talk the talk. The tools you will use to cut your metal are commonly referred to as either a *hardie*, or a *set*. What differentiates them is simply the direction from which you will cut the metal (either from the top or the bottom). They also look quite different from each other, but someone could have named them top hardie and bottom hardie, or top set and bottom set. Perhaps whoever named them was in a creative mood that day? Confused? Let's sort them out!

Set cutting tools are used to cut either cold or hot metal from above. They actually look like little hammers, but they are not intended to be swung. In the picture, you will note that one end is tapered, while the opposing end is flat. The smith will hold the tapered end atop the metal to be cut and then sharply strike the flat end with a hammer. This will cut the metal. Hot metal can be cut with a more acutely angled cutting edge. This is nicely accomplished with the narrower set, or *hot set*. In case you have not guessed, the job of the cold set is to cut through cold metal.

Essential First Tools Hardies and Sets

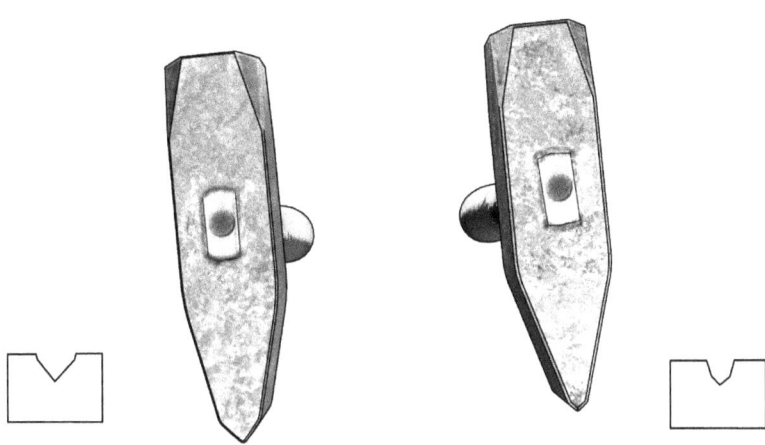

Hot and Cold Set

The hot and cold hardies are similar in design to the set tools, as the hot hardie is thinner and more tapered than the cold hardie. The difference when using these tools is that the hardie is seated into the hardie hole of your anvil. This will allow you to place the metal on top of the hardie tool, thereby cutting it from the bottom when you strike it with your hammer.

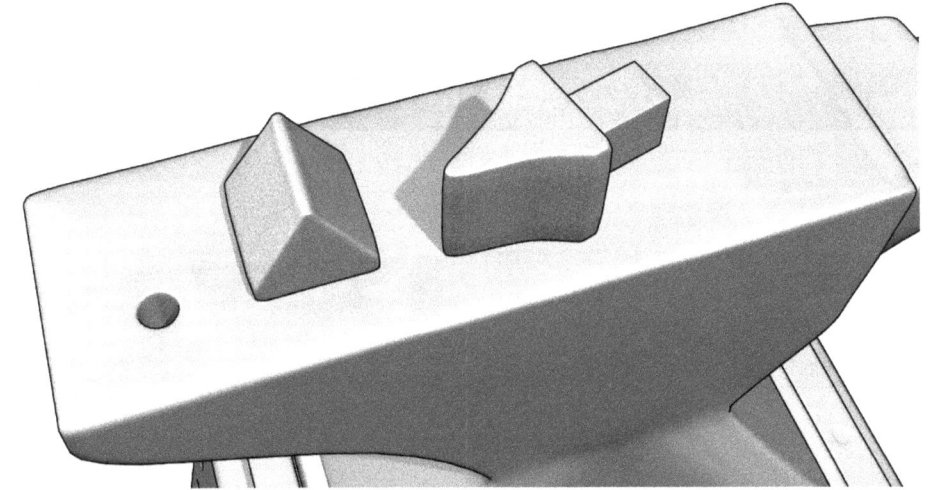

Hot and Cold Hardie

A set and a hardie can also be perfectly matched to work in conjunction with each other. As illustrated below, a set can be angled on one side and the hardie on the other, thus creating a sharp cutting tool. The hardie can also include a hollow, as pictured, for ease in cutting rounded pieces of metal.

If you're really comfortable with your aim, a cold chisel can be used in the same fashion as a set. The only real difference is that there is no handle to hold on the chisel. This method requires a confident strike of the hammer to avoid injury to the steadying hand.

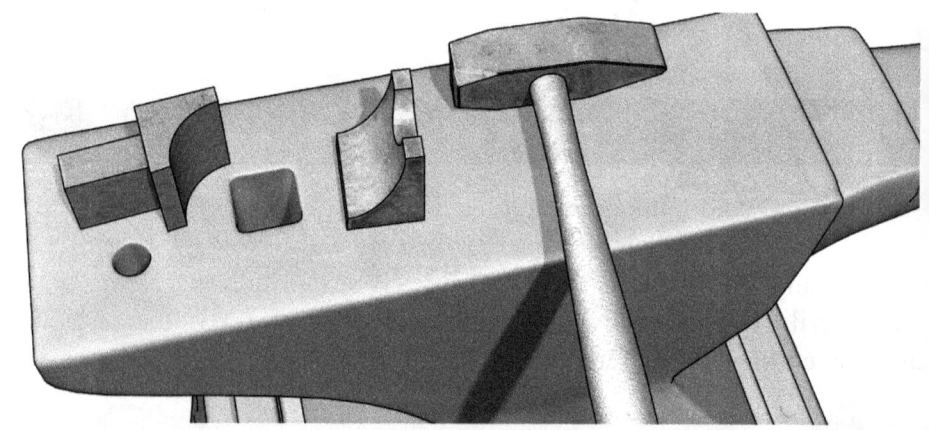

Cut Off Set

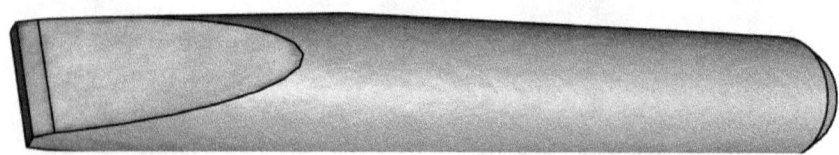

Cold Chisel

Fire Tools

Fire tools are used to help you command your fire. Simplistic in their design, these tools are good candidates to be the first tools that you can craft for yourself. The most streamlined of the fire tools is the poker. The poker is used to push coke and coal into the fire as needed. The end does not need to be sharp because you won't be stabbing your fire or impaling anything. You might find it helpful to bend the this end a bit to enable you to gather more material with fewer movements. The opposing end can be curled into a loop for you to use as a handle. This will also serve as a handy hook for hanging your poker out of the way when not in use. You will *not* want to leave it lying in the fire!

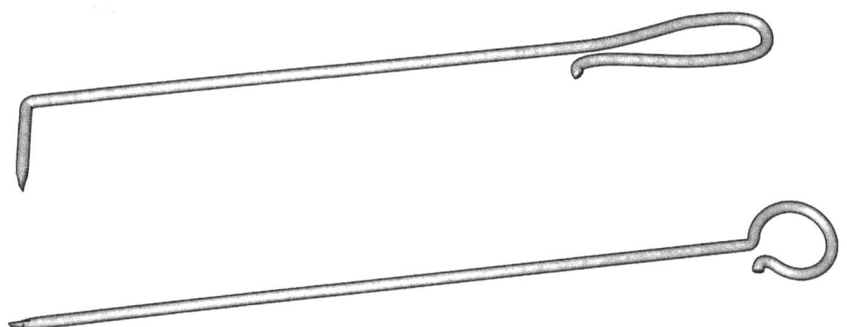

Pokers

A fire rake is similar to a poker, but will be used to pull rather than push material into your fire. The most commonly used "rake" does not look like a typical rake. One end is flattened and bent at a right angle to the shaft of the tool. The other end, as with the poker, can be bent into a hook or handle. A more intricate rake actually looks more like a 2 or 3 pronged garden trawl. Again, the opposing end of the tool can be formed into a hook for ease of use and storage.

Rakes

Another common smithing tool is the slice. This tool consists of a length of metal handle riveted to a flattened piece of metal. A slice basically resembles a flat spatula. The slice can be quite useful for lifting and moving coke and coal around the fire when using a rake would be impractical.

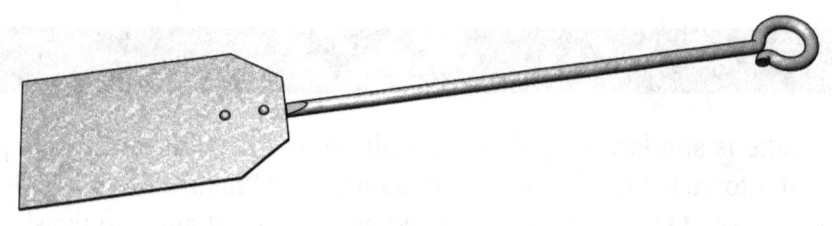

Slice

A small standard shovel should also be kept close to the forge for adding more coal to the fire when needed. This can be any small, store-bought shovel, but if you plan on making your own tools, why not start with a shovel?

Shovel

Any of the above mentioned fire tools can be fashioned from a 3/8 or 1/2 inch round steel rod with an overall length that will allow you to work a comfortable distance from the heat of the fire. The length will really depend upon the size of the forge, but 24-30 inches is within a standard range for these tools. The coal shovel is the only piece that will not need to have a long handle.

As was briefly mentioned in the chapter about safety, a professional blacksmith sprinkler should always be kept with your fire tools. The sprinkler can be made from any sized can and numerous holes should be punched into the bottom. A handle can then be riveted to the can. A handle that wraps around the sprinkler will be strongest. If you use a bolt to keep the handle wrapped around the can, it will be easier to replace the sprinkler if it should become worn and rusty.

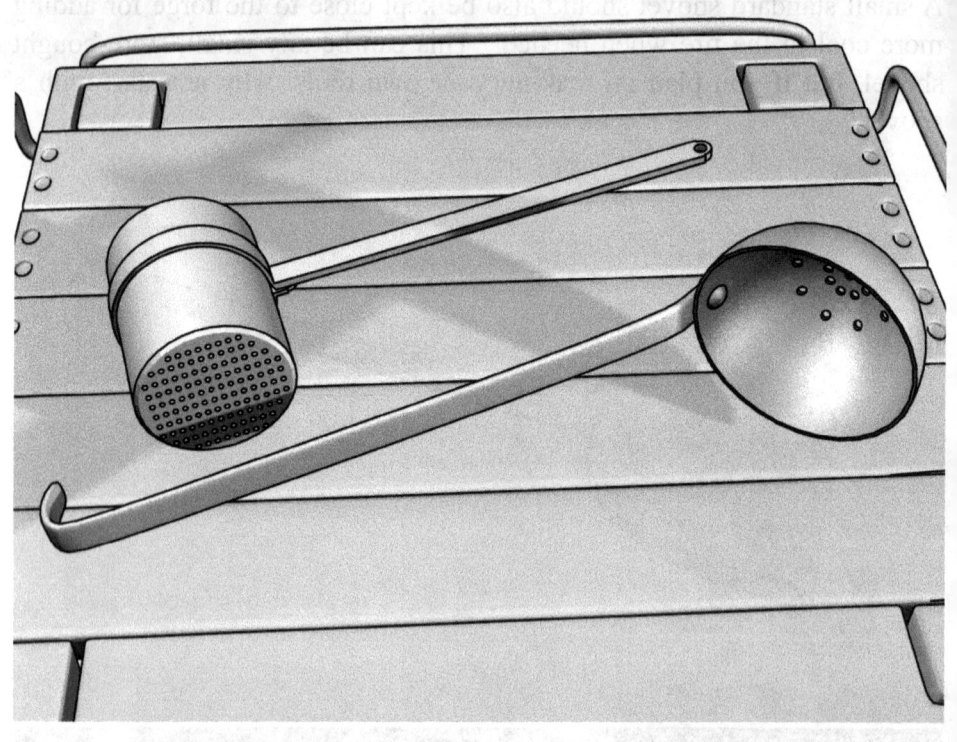

Watering Can

Brushes

You should have both wire and soft bristle brushes in your shop because both will be useful. When heated metal is withdrawn from the fire, you may notice that it is covered with scales and particles. Much of this can be removed by knocking the metal against your anvil, but a brisk rub down with a wire brush will do a better job removing the debris. A soft brush can be used to clean the anvil after you are done hammering.

Tongs

Tongs are a necessity in the blacksmith shop, especially if you choose to smith without gloves. Although longer pieces of metal can be kept cool by an occasional splash of water from your sprinkler, this will not be possible for small bits of metal. Most smiths prefer to work on the end of a long piece of metal, keeping the project long until they are ready to cut off the unwanted length. Even if you try this method, you will eventually need the assistance that tongs offer and, as with hammers, you will probably accumulate an assortment of tongs for various uses. As you identify your own needs you can craft specialized tongs made especially for your unique needs.

Some modern pliers may be used as smith tongs, depending upon the size of the project and the forge. Pliers are typically not as long as blacksmith tongs. Therefore, they will probably not work very well for smaller projects as they may put your hand too close to the heat of the fire. If you can use pliers, the hand vise style will work very well and should maintain a tight grip on the metal.

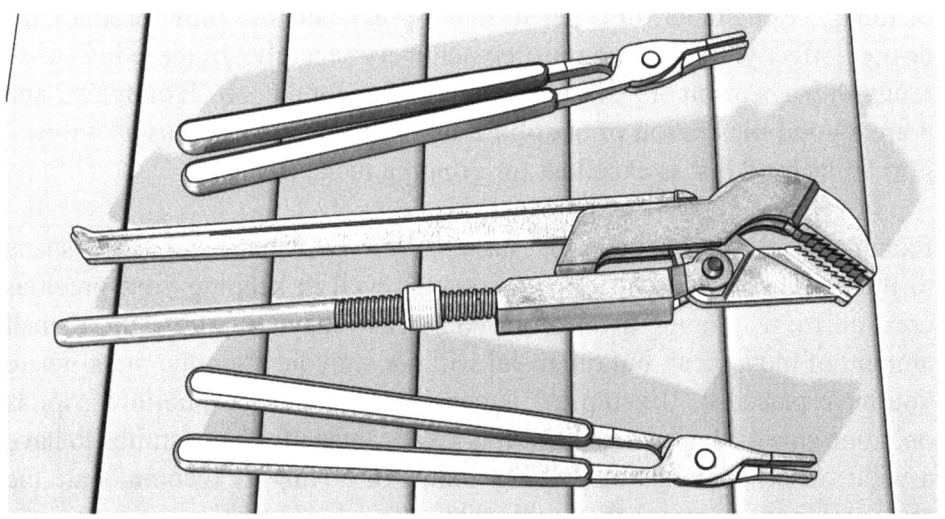

Pliers

Blacksmith tongs are very similar to pliers, but they have much longer handles that allow you to keep a safe distance from the fire. The actual length of the tongs will vary, but typically they don't extend past 18 inches. As a rule of thumb, the longer the length of the handles, the greater the leverage and grip will be at the jaws.

All tongs have the same basic design. The pivot of the tongs is usually nothing more than a rivet holding them together and the handles are forged to taper. The gripping handles are often rounded but, no matter their shape, they should never touch. Tongs can be forged into countless shapes to grip customized creations. A nice addition to a set of tongs is a "locking ring" (or coupler) which can be slid toward the handles to lock the jaws onto the metal. This will ensure a tight grip on the project and can save your hand from fatigue. (See example 1 in the Tongs illustration.)

The best grip is achieved when the metal is aligned with the tongs, but there may be times when you need to turn your creation to shape it. In these situations you may find that the tongs are in your way and only a cross grip will do. Beware because this positioning is very unstable, allowing room for your creation to spin, especially if you are working on a rod-shaped piece of metal. Remember, in order to make every hammer blow produce the desired effect your creation must be held very securely. In these instances, tongs with a bent-bit or "side tongs" are perfect for the job. Examples 2 and 3 are a good illustration of bent-bit tongs. Example 2 also has a "hollow" grip on its head that is excellent for gripping metal rods.

Example 4 illustrates tongs that have hollow grips but have a box shape to them. These types of tongs work very well in keeping your precious creation from spinning away from you. They may still allow for a small amount of movement, but the metal will not stray far from the mark where you have placed it. Example 5 is considered a semi-box hollow grip, as only one side of the jaw has lips. It is an advantage for blacksmiths to have a wide variety of tongs of varying widths of boxing to accommodate the many different creations they will make.

Essential First Tools Tongs

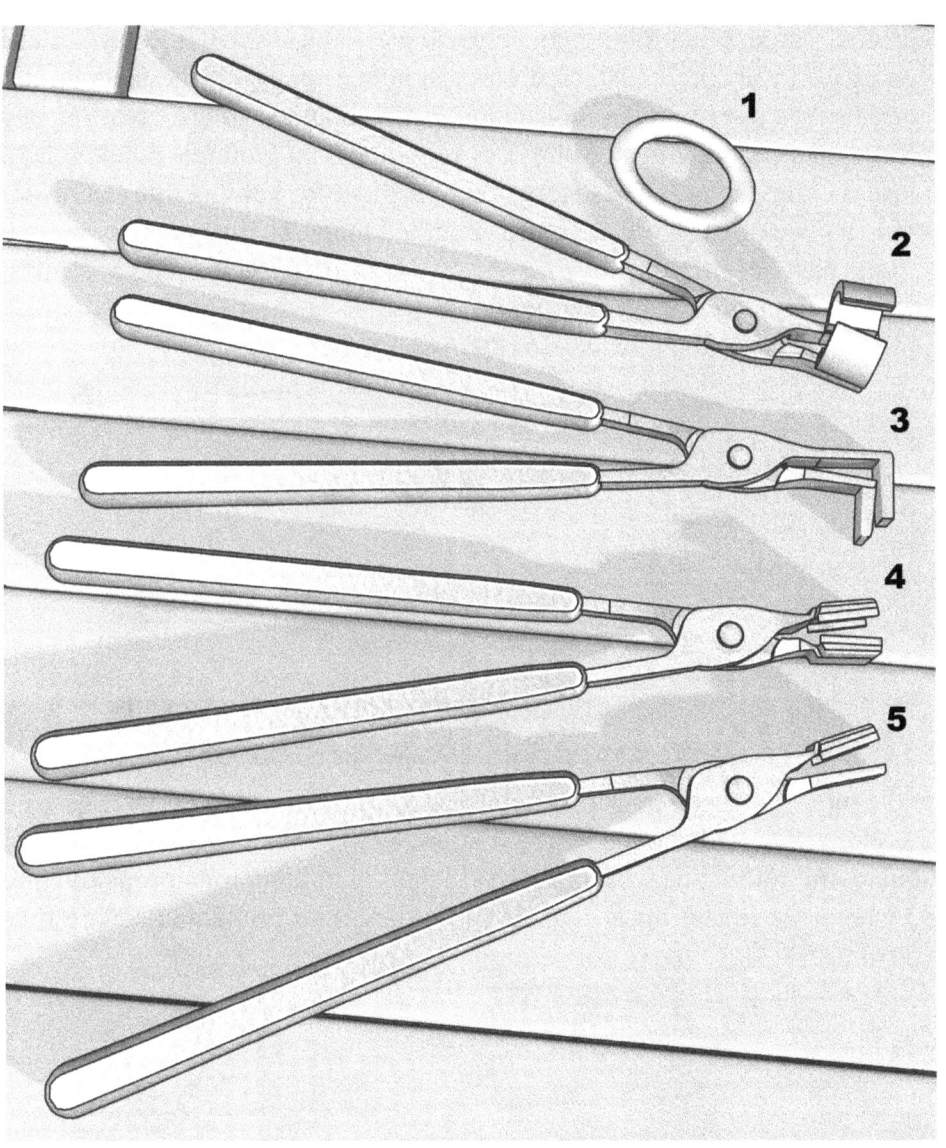

Tongs

General purpose blacksmithing tongs have flat jaws. If the jaws close completely, they are considered close-mouthed tongs. If the jaws do not completely close, they are considered open-mouthed tongs. Not exactly rocket science, but it is nice and easy to remember. Example 6 below is of close-mouthed tongs. These tongs are excellent for holding thin flat pieces while they are being forged. Open-mouthed tongs, as illustrated by picture 7, will best hold flat pieces that are a ¼ inch or thicker. You will most likely accumulate many open-mouthed tongs, as the degree of openness will depend upon the thickness of the metal to be held.

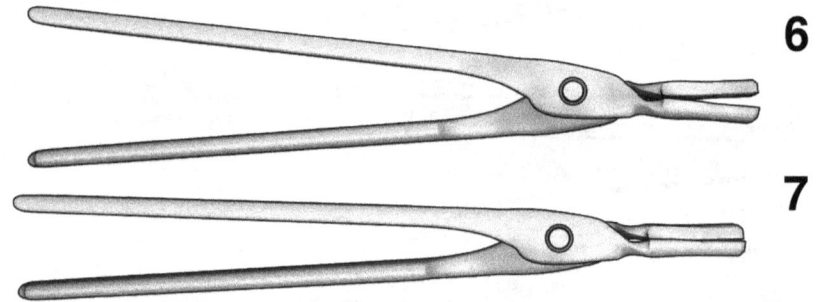

Open and Close-Mouthed Tongs

If the piece you need to grip is rod-shaped, you will need to use tongs with a hollow jaw, as illustrated below (picture 8). The jaws may be rounded or square, but round tips will only be sufficient in holding round rods (picture 9) whereas a square tip is capable of holding both round and square rods (picture 10).

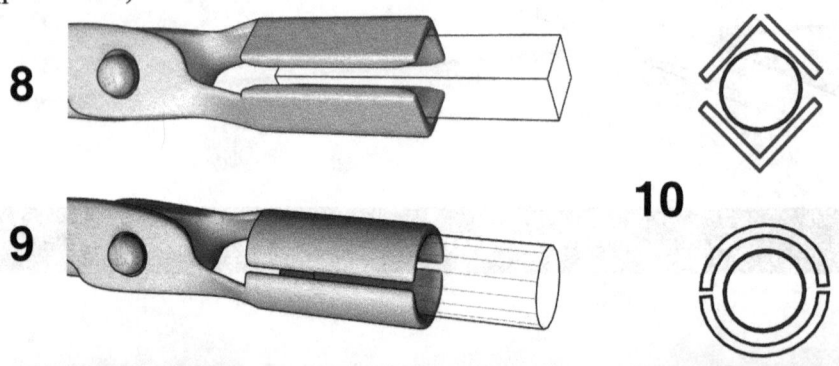

Tongs with Hollow Jaws

Just as you have different sized tools for driving screws or loosening bolts, you will need several tongs with varying jaw sizes. Hollow-tipped tongs will only accommodate parallel rods. Therefore, if you have a piece of metal with an enlarged end, such as one that has had a head forged onto a rod, bolt tongs (picture 11) can be altered to suit your needs.

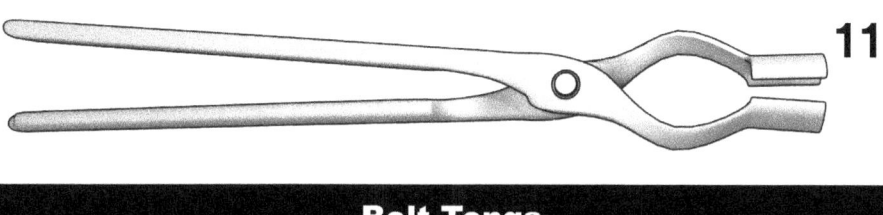

Bolt Tongs

There are also tongs specialized for riveting. Oh, sure, you could probably use another type of tong to hold the small rivets in the fire, but there is a reason that customized tongs exist for riveting work. Because of their small size, rivets will cool much quicker than a larger piece of heated metal would. Example 12 below shows tongs that are designed to match the diameter of a rivet. By positioning a rivet prior to heating, it can be lifted out of the fire and immediately seated where you need it. In doing so, you will not lose any heat while fumbling around with the small piece.

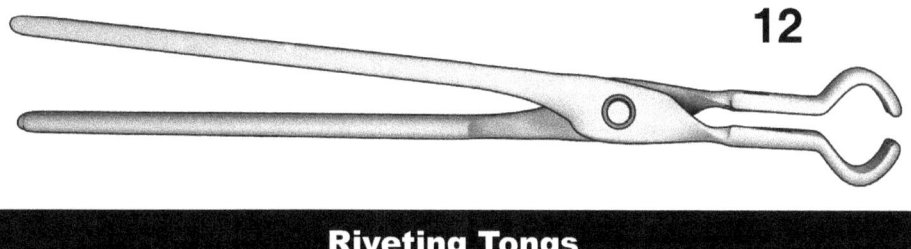

Riveting Tongs

There are also pick-up tongs (picture 13), whose sole purpose is just as their name suggests – they pick stuff up! They have a more "springy" action than standard tongs and are used to transfer a piece of metal to a vise or other tongs, but are not intended to hold the metal while forging. You will also notice that they have a couple of openings in the jaws, which will allow you to securely pick up metal of varying sizes.

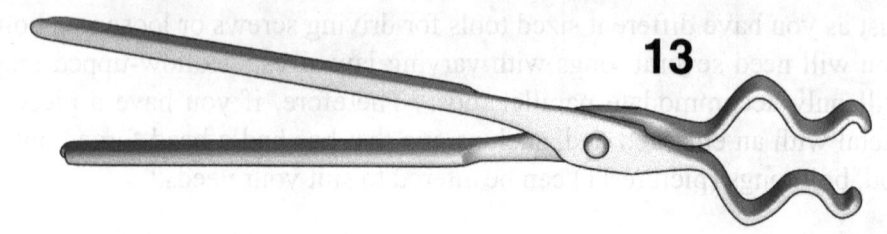

Pick-up Tongs

Measuring Tools

Measure twice; cut once. That is the golden rule of any project, right? Surprisingly, a blacksmith does very little measuring. Sizes and angles are generally checked by eye or by direct comparison rather than by taking measurements. Many blacksmiths will forge their own calipers to measure with. The example below shows a caliper with dual measuring capabilities. Such pieces are typically made with a handle to keep the smith's hands clear of the heated metal. They are simply crafted with small pieces of metal held together with tight rivets, which can be moved to precise points for measuring.

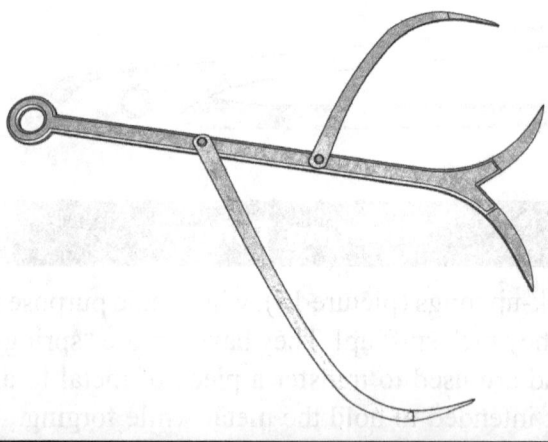

Caliper

A right angle is also handy to have around the shop. They are easily fashioned from strips of metal that are welded or riveted together. Again, a handle can be fastened to keep hands clear of the heat. Larger right angles may need a diagonal brace placed between the sides for stability.

Right Angle

For measuring inches of heated steel, the blacksmith may employ an engineer's ruler for the job, but this is often not practical. Because metal is a great conductor of heat, the metal ruler will quickly become too hot to handle. Another potential problem is that rulers are typically made of steel. The heating of the ruler during measurement and the cooling process afterwards can actually change the temper of the instrument, thus leaving it soft and bendable. A ruler made of brass may be a good alternative because it will not structurally react to the heat, but you will still have to contend with the ruler becoming too hot to handle. So what's a smith to do?

A traveler is a measuring tool that was used many years ago to measure steel for wagon wheels. Although not often used today, it still enjoys a place in many smith shops. The tool is a metal wheel attached to a handle. The wheel is marked, either by a notch or hole, at the edge of the outer wheel. To use the traveler, start the wheel with the hole at your starting measuring point and then roll it along the material you are measuring. You would then count the rotations of the little hole to determine the measurement. Not very technologically advanced, but it works.

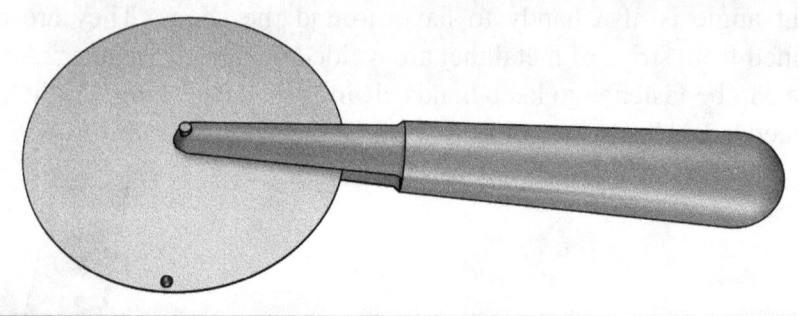

Traveler

The blacksmith is a fortunate craftsman indeed, as he or she will be able to make whatever tools may be needed. There is no need to wait until some manufacturer discovers the next new tool to make your job easier. Your tools can be made for your own specialty line of crafts. Even if you need a special tool for one project, you can always re-craft it for other projects, changing it as you see fit. Being able to do this is just one more satisfying aspect of the craft you have chosen.

Essential First Tools Measurment Tools

50 Micrometers **100 Micrometers**

6

Choosing Your Metal

Choosing Your Metal

You are now ready to select your metal, but what type do you choose? How do you choose? Any good blacksmith will collect scrap metal, but how can you tell what type of metal you actually have? Do you need a degree in chemistry? Now that you are riddled with questions, let's see if we can sort it out.

Iron and steel are used for almost all blacksmithing. No other metals can provide the same results from heating and reshaping. Steel is a derivative of iron, but the base metal for both is iron ore. Iron and steel are obtained from the ore through the heating of the base metal. When ancient man discovered iron could be extracted from ore, he needed to be quite creative in finding ways to sufficiently heat the fire to complete the task. He would have needed to go to great lengths to build up the heat by trying something like directing the wind into a canal to blow across a fire pit. Those early fires, fueled by wood, gave way to charcoal or coal fires and the invention of the bellows further eliminated man's dependence on Mother Nature's moods. After heating the ore, the fire was raked out and the infant iron was collected. Because of inexperience, the quality of the iron could not be controlled and thus was often plagued with impurities that could leave the metal weak and brittle.

Today, iron is extracted in huge blast furnaces that produce a material called pig iron. This initial iron contains small, but numerous, amounts of impurities. Although pig iron is 95 percent iron and up to 4 percent carbon, the remaining 1 percent can contain various levels of silicon, sulfur, phosphorous, and manganese. So, what does all of that mean? The impurities will have a considerable effect on the quality and characteristics of the metal. Although a blacksmith does not need to have a degree in chemistry, it is important to understand the constitution of iron. Differences

Pure, carbon-free iron is soft and weak. The more carbon content a metal has, the harder and stronger it will be. Carbon content of 0.9 percent creates alloys of steel. Carbon content of approximately 2.0 to 3.3 percent creates a "semi-steel" referred to as white iron. Gray cast iron is created when carbon is excessive and insoluble in solid iron and it remains present as finely dispersed graphite flakes, which act as "stress raisers." When cracks form within the iron, the flakes actually encourage the cracks to spread and worsen. In 1948, a new process of creating ductile (metal that can be flattened or "drawn out" without losing its strength) cast iron, produced graphite flakes that took on a spherical shape. The newly shaped graphite actually prevented the cracks from spreading. The two graphite shapes are pictured below:

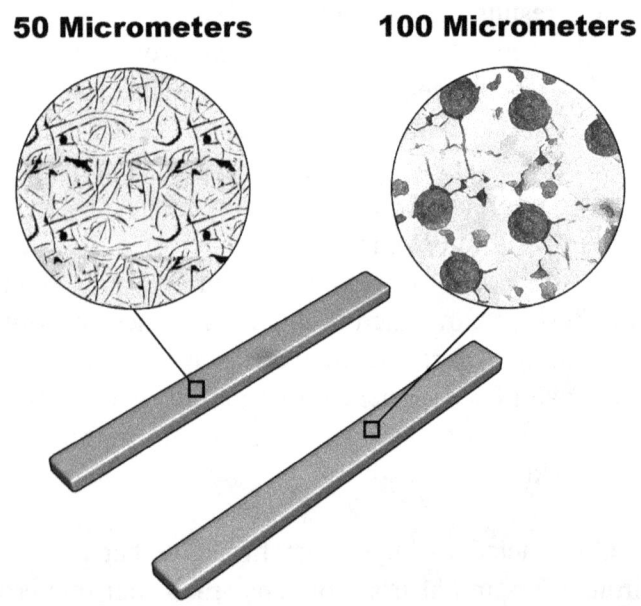

Graphite Flakes and Graphite Spheres

Cast Iron

Cast iron is created by re-melting pig iron and pouring it into molds. The quality and strength of cast iron is controlled by varying the chemical make-up of the metal and by cooling it either rapidly or slowly, depending on the desired results. Cast iron is excellent for making machine parts or anything else where the weight and bulk of the metal is desirable. It is an extremely hard metal but is not of much use in the smith shop as it cannot be altered by heating and hammering. Ductile cast iron can be accomplished through additional processing, but it is still not the metal of choice for blacksmithing. Ductile cast iron can be machined successfully, but is generally used for parts that are turned and formed with cutting tools.

Wrought Iron

Wrought iron is the favorite metal among blacksmiths as it is the toughest, most ductile, and most malleable form of iron. After pig iron is produced, it is further heated and rolled to reduce the carbon and to rid the iron of most of its impurities. The result is wrought iron. When produced in strips, wrought iron has fibrous characteristics which make it particularly responsive to shaping with a hammer. It is also quite resistant to corrosion. Light rust may form on the metal, but the coating serves to protect it, thus preventing further corrosion. Rest assured that a creation made from wrought iron will last for years to come.

Now for the bad news – wrought iron is not readily available today and has been generally replaced by mild steel (iron that contains a small amount of carbon). Mild steel is superior for structural work, machining and general engineering, but it is less desirable for smithing. Oh, mild steel will be perfectly acceptable for most of the blacksmith's crafts, but it doesn't work well in delicate projects as it is more difficult to weld with the smith's methods.

There is still a small amount of wrought iron produced, so you may be fortunate to find a supply. For the most part, however, the modern blacksmith will probably have to settle for mild steel. If you come across some old wrought iron railings while searching through scrap metal - grab them! They are a blacksmith's treasure!

The amount of carbon in mild steel does not affect the hardness of the metal. A blacksmith can alter the hardness of many metals by heating and cooling them. This method is referred to as *tempering,* which will be discussed in a subsequent chapter. However, heating and cooling mild steel will not appreciably affect its hardness or softness. The best steel for the blacksmith to use for making tools will have a carbon content of approximately 2 percent and is referred to as high carbon or "tool steel". Two percent carbon steel responds well to tempering and can be made harder by heating and quenching the metal and can be made softer through another process. Depending upon your creation, you will be able to control how hard or soft the metal will need to be.

Although there are many different steels available today, it is still recommended that the smith use only ordinary high carbon steel to craft tools. You will find that many of the "specialty" steels have other metals added to the steel to produce very specific qualities for their intended uses. These metals often require precise heating treatments that are attained through the use of specialized equipment that the typical blacksmith does not maintain.

Alloy Steels

The term "alloy" applies to a metal that contains a mixture of two or more metals. Steel is sometimes referred to as an alloy of iron and carbon, but this statement is really not accurate, as carbon is not a metal. With modern technology, the amounts of metal that are added to steel can be precisely controlled and the additions can produce considerable changes in the characteristics of the metal.

For example, 18 percent chromium and 8 percent nickel added to steel will dramatically increase the metal's resistance to corrosion, which is more commonly called stainless steel. High speed steel is produced by adding one or more of several different metals, which creates a very hard material that will stand up when heated. This particular steel is excellent for high speed cutting blades. The specialty metal choices that are available today seem almost endless, but try not to rack your brain too much when trying to choose. It is best to stick to good old high carbon steel that is free from any alloy additives.

So, what exactly is it?

Okay, you're wandering around a flea market looking for scrap metal and you come across what looks like some suitable material. How can you tell how much carbon it contains and if it would be suitable for smithing? There is actually a very easy test you can perform to find out! Using your grinding wheel, hold the metal against the spinning wheel and watch the sparks fly! You will look at the trail produced by the sparks and the amount of "stars" it contains. Low carbon steel will produce a long spark trail with very few stars. If the metal contains a medium amount of carbon, the trail will be shorter and produce more stars. High carbon steel will produce short bursts of sparks with numerous stars. There are also some depictions of a couple of other common materials you will come across in your search for the perfect metal.

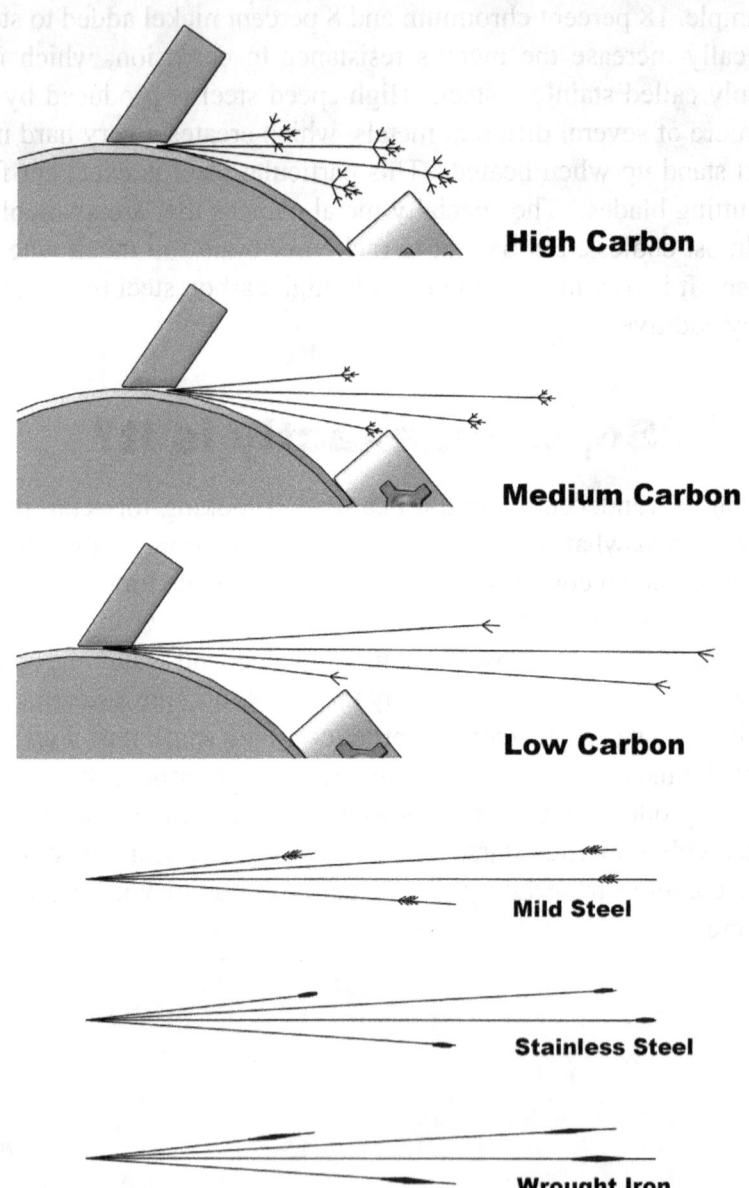

Sparks

If you don't want to purchase all kinds of metal only to get it back to your shop and have it fail the grinding wheel test, you can also use a hand-held grinding wheel to perform the same test. This way, you can check your metal and know if it is right for you before purchasing it. Keep in mind when performing this test that you must apply the same amount of pressure to the grinding wheel with each sample so that the results can be accurately gauged. As with anything else, it may take a bit of practice before you can expertly interpret the sparks.

During your scrap searches, keep in mind that any metal that has been used to make a tool has the possibility to be reworked into another tool, even if the original appears to have outlived its usefulness. Good sources for high carbon steel are vehicle springs, leaf springs, and even coil springs. You may not find as many uses for coil springs, but they can be straightened and cut to make tools. Any ornamental or fancy metal work that you suspect is older should be grabbed, as it is most likely highly prized wrought iron.

Of course, you can also purchase new mild steel, which may be the way to go if you are buying in larger quantities, though some of the romance of blacksmithing may be lost when working "new" metal. You will gain a great deal of satisfaction from searching for and finding that "perfect" piece of metal and crafting it into something beautiful. Upon its completion, your creation will already have a history, and you will have an interesting story.

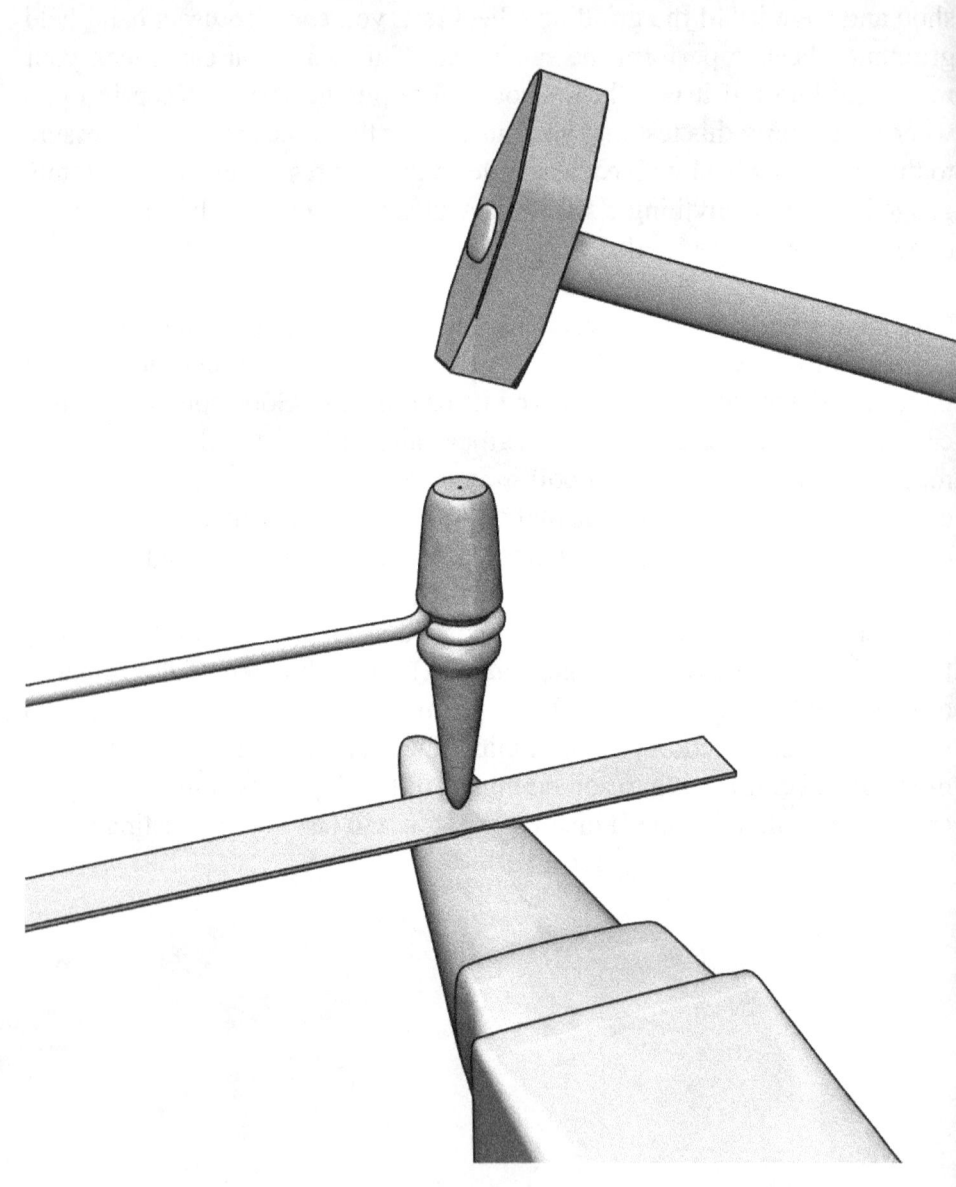

7

Techniques for the Beginner Smith

Techniques for the Beginner Smith

By now you have set up your shop and have all of the equipment you need to begin blacksmithing. Your forge is built and stands at the ready. Perhaps you have even practiced building your fire – good for you! You searched for, and found, the perfect, most beautiful sounding anvil. Piled in the corner of your shop is a sufficient beginning supply of scrap metal, along with an ample amount of animal hooves and horns (um…we'll get to that later). All of your equipment is right where it should be and the quenching bath sits quietly, waiting for that first sizzle of hot steel you will plunge into it. You have even gone out and purchased your first hammer. So, you're ready to dive right in to start making all of the creations you have yet only dreamed about. Well, cool your heels, you're not there yet!

There are basic smithing techniques you must master before you attempt to create even the simplest of tools. You will technically be blacksmithing, but the exercises presented here are for you to practice the different techniques of shaping heated steel. Once you have mastered these techniques, you will be bound only by your own imagination. A wonderful aspect of blacksmithing is that once you have practiced on your scrap metal, the metal can be reheated and used again, so don't think that you're wasting your supply. It is also freeing to know that if you make a mistake, you can reheat the metal and start over.

There are some general thoughts to keep in mind when heating your metal. A steady blast of air should be maintained to sustain the fire's intensity. You will need to experiment a bit to determine how long a piece of metal should remain in the fire to reach its desired temperature (the color of heated metal will be mentioned throughout this chapter). Also, common sense plays a roll here, as thinner pieces of metal will heat more quickly than thicker ones.

If you are heating a tapered piece of metal, where the end is both thick and thin, you will need to manipulate the metal in the fire to avoid overheating the tip. The thicker portion of the metal will, obviously, take a bit longer to absorb heat.

Hammering

Material and equipment: A hammer (duh), your own muscle power

You may think that swinging a hammer doesn't take much thought, but if your forging is not properly executed exhaustion will quickly set in. There are techniques to swinging a hammer so that you will get the maximum effectiveness, while expending the least amount of energy. Like all of the techniques presented here, hammering must be practiced until it becomes an automatic movement.

The first step is to make sure that you are properly standing at the anvil. Your legs should be comfortably spread apart to ensure balance. One foot should be slid back slightly while the other is moved forward so that it is actually under the overhang of the anvil. Now, you'll want to keep a close eye on each hammer blow (literally), so you should hold your head close to your work. Your head should be bent over the anvil and your work, but kept off to the side so you don't hit yourself with the hammer. This may sound a bit silly, but it is actually easier than you may think to whack yourself.

The hammer's beginning position during a swing will be above your head and over your shoulder. If you're not paying attention, you could clip your head with the hammer on its way down to the anvil. Also, there is a real danger in having the hammer recoil from a misplaced blow. It could come back up and hit you in the face, so be careful!

You should never use only your shoulder to execute a swing. Hold your shoulder steady during a swing and allow the other muscles and joints of your arm to take up the slack. Try to transfer the energy from your shoulder throughout the rest of your arm during the execution of a blow.

Techniques for the Beginner Smith Hammering

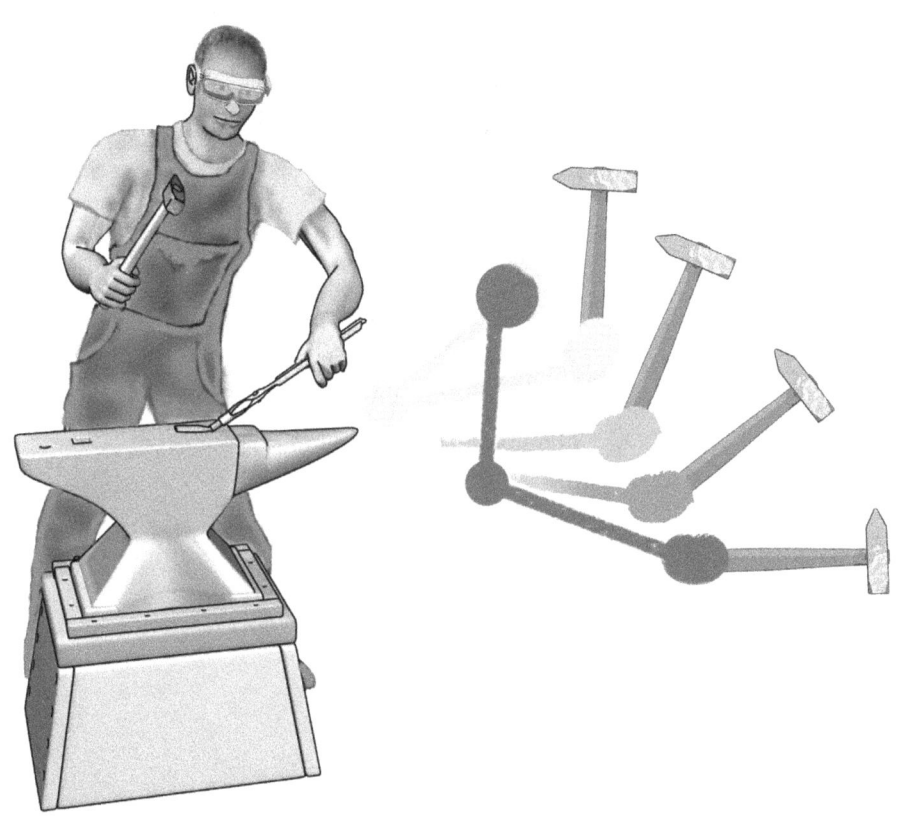

Hammering

You should always wear your goggles when you are forging. As the hammer hits the heated metal, particles of steel and oxide scales will be released and introduced into the air. In addition, very small debris, such as cinders and ash dust from the fire, will be flying through the air. It just makes sense to protect your eyes, especially because your face will be so close to the source of the flying debris. When hammering small or delicate objects, the beginning position of the hammer should be at about shoulder height. The hammer should be cradled loosely in the fleshy web between your index finger and thumb. The rest of your fingers should be positioned in a line along the hammer's handle.

With the hammer angled backward in the cradle of your hand, you will create the first movement downward by pulling on the handle with your fingers. You will then tighten your grip as you follow an arching blow to meet the metal. This technique is also best when riveting small rivets or light-weight nails.

Hammering Small Objects

When practicing your hammering technique, it is best to do so with cold steel. Also, practice by holding the metal in your hand, vice grip pliers, or tongs to get used to actual forging. Practice with different hammers. The point is to practice until you are completely comfortable with executing hammer blows so that you can forge hot steel and accomplish the intended results. Many blacksmiths will fall into a rhythm when hammering, such as three blows to the metal and one to the anvil. The rhythm helps to keep each blow focused on where you want it to land. You will not want to simply pound away on your creation. Each hammer blow should land exactly where you had intended, while producing the result you had intended.

Cutting

Material and equipment: iron or mild steel, hardie or set tool, peen hammer

To practice basic cutting skills, it is recommended that you use round pieces of either iron or mild steel that measure from 1/4 to 3/8 inch in diameter. Try to use strips that measure about 18 to 24 inches long, rather than smaller pieces that would need to be handled with tongs. You'll want your focus to be on your cutting technique, not worrying about whether you may drop the heated metal. You will have much more control during the learning phase if you can hold the metal in your hand.

It is possible to cut metal while it is cold if you have the more acutely angled hardie or set tools available. If they are not available, heated metal will cut quite nicely with the thicker cutting tools. If you have both types of tools, you may want to try cutting cold and hot metal to note the differences. This will allow you to determine your preferred method. If you heat the metal for cutting, leave it in the fire until it heats to a red color. You'll want to check it often until you become proficient in judging how long the metal will need to remain in the fire.

Cutting metal with the hardie tool is actually easier than using a set. With the hardie tool mounted on the anvil, both of your hands will be free. In one hand you will hold the metal, and in the other, your peen hammer. If using a set, you will be short a hand, as the metal will need to be steadied while you hold the set tool in one hand and the peen hammer in the other. In most instances, when cutting with a set, you will need to employ the help of another person.

When using a hardie tool, hold the strip of metal over it at the point where you want the cut. Give the metal a good tap with the flat end of your peen hammer, as shown in the picture below. Do not hit the metal so hard that the cut occurs with one strike. If you do, your hammer will strike your hardie tool and you risk damaging the tool.

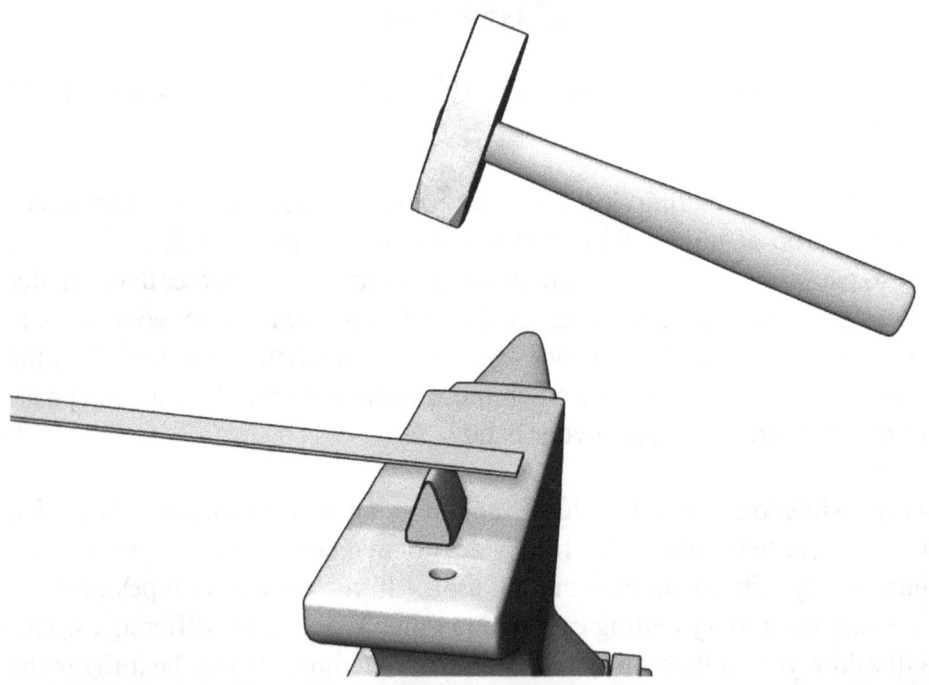

Using Hardie Tool

Once the metal has been "dented", turn the strip over and strike it in the same manner, without completing the cut. After the two strikes, the metal should have two opposing notches, as pictured below:

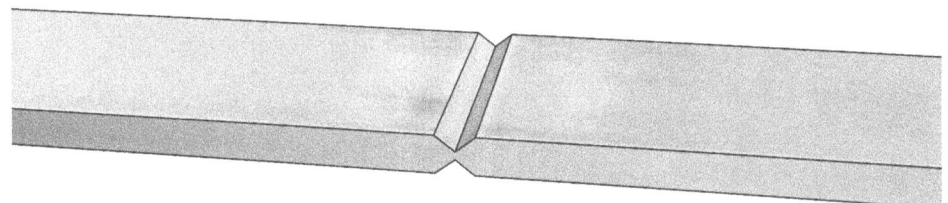

Notches

Once the metal is notched, you may be able to complete the cut by bending the metal by hand. You could also insert one end into a pritchel hole in the anvil and lever it apart. If a set tool is your preferred method of cutting, consider performing this task on a table, rather than the anvil. If you happen to cut completely through with one hammer strike, you again run the risk of damaging your tool if the set meets the anvil with enough force. You will notch both sides of the metal in the same manner you would if using a hardie tool. Complete separation can again be achieved by hand or with the help of a pritchel hole.

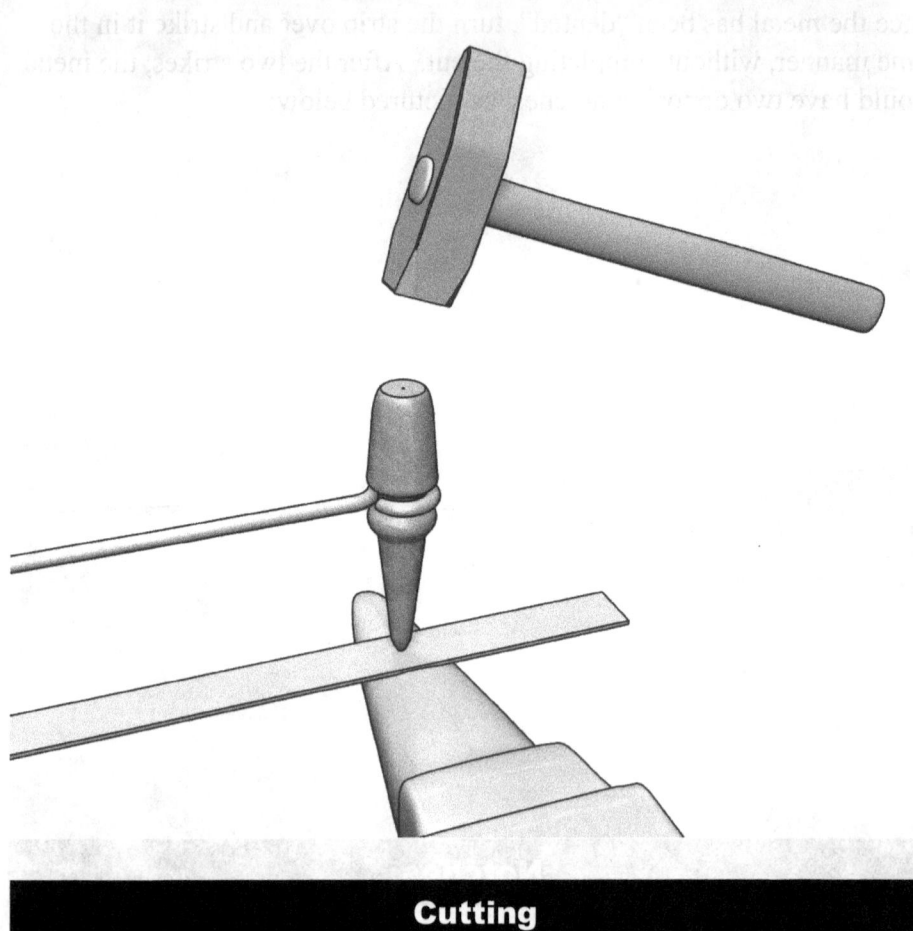

Cutting

Actually, you could just place the metal into a vice and use a hacksaw to cut it, but what fun would that be? Most blacksmiths prefer to cut metal by hand. Once you become more proficient in cutting, you will see that a couple of strikes with your hammer will be easier than taking the time to place the metal in a vice and sawing it in half.

Bending

Material and equipment: length of iron or mild steel, ball peen hammer, scroll iron or "fork" (optional)

Much of the work you will perform as a blacksmith will involve bending metal. As with all of the skills presented here, it will take practice to bend the metal without pinching it between the hammer and the anvil. Carefully placed blows will prevent the hammer from leaving unintentional marks in your creation.

You will need to heat the metal to a red color and ensure that the same amount of heat is achieved for the entire length that is to be bent. The color may vary slightly depending upon the metal's constitution, but a red glow should be satisfactory. You will not want it to get too bright so that it begins to turn golden or orange. If it becomes bright and sparkling, you have overheated it. Overheating the metal will "burn" it (yes, metal can burn), leaving it degraded and unsuitable for forging.

The beak of your anvil is ideal for bending metal and it is thoughtfully tapered for varying degrees of curling. When striking the metal, the blow should make contact to one side of the point of support (Picture 1). If you want to create a large curve, work the metal across the beak and alter the strike point by moving the metal (Picture 2). Remember to always hit it off-center so that you don't pinch the metal. As you become comfortable with the process, you can strike the metal toward the beak to close the curve. Avoid hitting the metal too hard against the beak for previously stated reasons that won't be mentioned again.

When shaping metal into a curve, you will most likely find that the metal will need to be reheated during the process. You will do this when the metal has lost its red glow. If you continue to hammer after the metal has gone black, progress will slow to a crawl. You will also notice that more forceful hammer blows will be required to get the metal to bend, which will leave undesirable marks behind. Keep a good eye on the color of the metal as you forge and take the time to reheat it when necessary.

Some smiths prefer to curl metal over the edge of the anvil's face rather than the beak (Picture 3). As with curving metal over the beak, the strip is forwarded over the face to progress the curl. Another method of curving metal does not require a hammer and is achieved with the use of a scroll tool which can be mounted in the hardie hole or held tightly in a vice (Picture 4). The shaping is done by levering the strip of metal back a little at a time from the end and then progressing down the rod. This type of bending works much better with a piece of flat metal, as it can leave marks in round pieces. It is best to stick to hammering if you're using round pieces of metal.

When making your own tools, you will want many of them (particularly the fire tools) to have an eye on the end for ease of use and so that you can hang them up when they are not in use. First, decide on the size of the eye you want to make. You will create a sharp bend in the metal, allowing approximately three times the length for the intended diameter of the eye (Picture 5). Utilize a narrow part of the anvil's beak that is slightly smaller than the eye is to be. Hold the rod so that the end can be moved progressively as the curl is created. Start with the end and work up the piece of metal (Picture 6). You will continue to move the metal and strike it with the hammer until the eye hook is closed (Picture 7).

Don't get discouraged if your round eye hook does not end up looking like an actual circle. You can reheat it and hammer around the beak to define the shape, or just leave it as you initially made it. The purpose is to have a hook on the end. It does not have to look perfect. As you hone your skills, you might even be amused by seeing your well used tool with the misshapen eye hook hanging in front of you. It can serve as a reminder as to how far you have come with your blacksmithing techniques.

Techniques for the Beginner Smith Bending

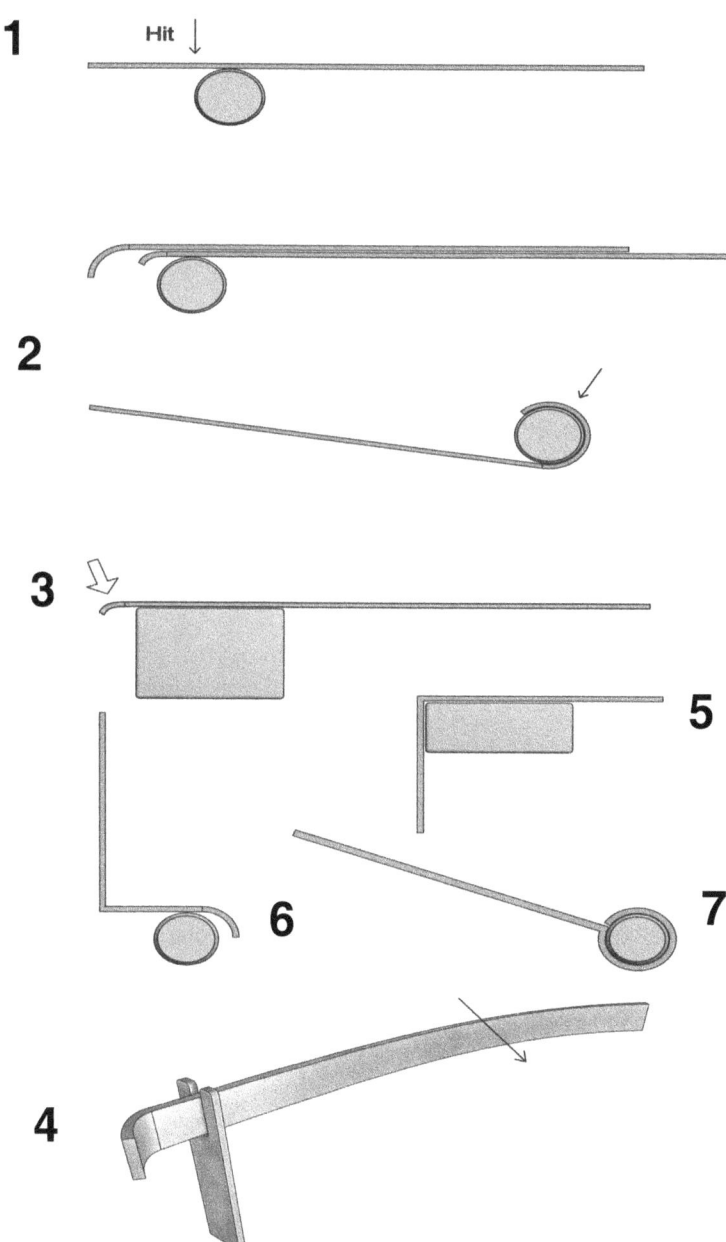

Bending

Drawing Out

Material and equipment: length of iron or mild steel, ball peen hammer, fuller (optional), flatter (also optional)

First, let's introduce the blacksmith fuller and flatter as they have not previously been presented. A fuller consists of matching pairs of curved tools (pictured below). With the bottom fuller seated in the hardie hole, the top fuller is held over the metal and struck with a hammer. They are generally used to pinch and hollow metal that is worked between them, but they can also be used to draw out metal, which is why they are being introduced in this chapter.

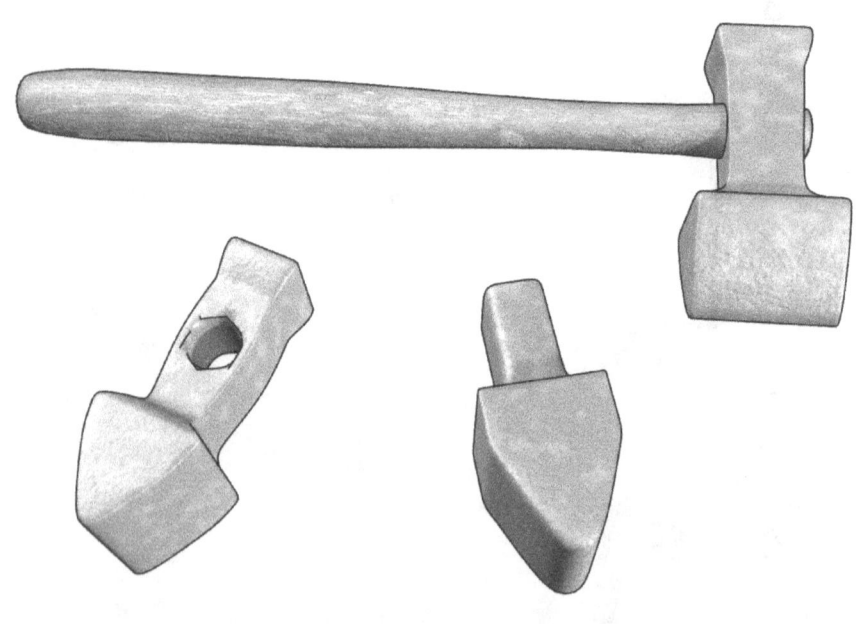

Top Fuller

When hammering metal, tool marks can be left behind that mar the surface. As you progress in your blacksmithing skills, you will no doubt be able to limit the marks left on your creations. However, to obtain a true flat surface, there is no shame in employing the help of a flattening tool (pictured below). The flatter is placed over hot metal and struck with a hammer, which produces a smooth, flat surface and evenly distributes the hot metal.

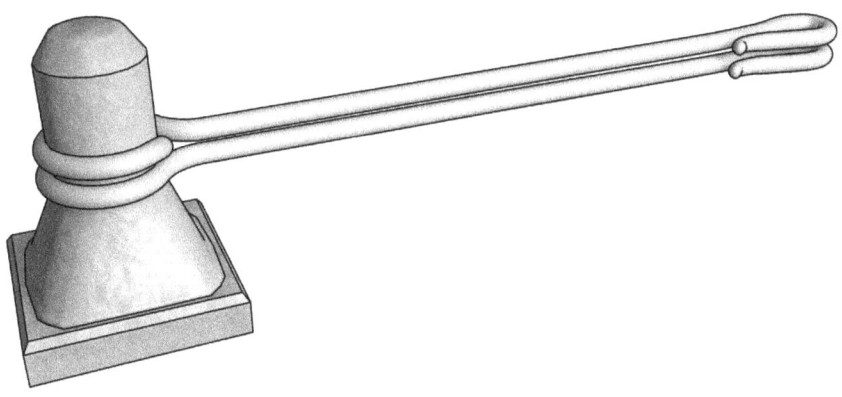

Flatter

Drawing out metal (also called "drawing down") is the term that refers to the process of tapering a rod, lengthening a piece of metal, or flattening it. For example, a square rod can be tapered into a flattened wedge, a square spike, or a rounded spike. Basically, the process of drawing out will make any piece of metal thinner and longer. Keep in mind when you are drawing out metal, that it is impossible to compact an area of the material without making it longer or wider in another direction. That makes perfect sense, doesn't it?

Most tapering should be performed on metal that is heated to a bright red color. However, when you complete the final shaping it is best that the metal be dull red in color. Be careful, as you find the need to reheat the metal during tapering that you do not allow the thinning tip to overheat. You will also find that metal heated to a bright red color will form black scales on its surface. Don't worry, most of the scales will be knocked off during hammering and will not compromise the integrity of the metal. Be sure to brush the scales from your anvil often to maintain a smooth working surface.

If you are using the anvil's beak as a base in drawing out the metal, be sure to strike the metal directly on the top (Picture 8). The curve of the beak will assist you in forcing the metal towards the end, as well as thinning it. The tapering process should start at the end of the rod and then proceed back (Picture 9). If the taper is dramatic, most of the hammering will be done towards the end of the rod and continue back with a less heavy hand. After tapering the length you desire, turn the rod over and do the same on the other side. Be sure to move the rod to accommodate the blows of the hammer, keeping a hard surface underneath it, which will effectively "pinch" the metal. (*Now* you get to pinch the metal.)

An alternative to using the beak for tapering is to utilize the edge of the anvil's face (Picture 10). The rod is hammered against the edge, effectively pinching and lengthening the metal. The process will create notches in the metal and you will also notice that the rod will begin to curl as you progress. But, not to worry. Turn the rod over and repeat the process and you will see some straightening as the opposing side is "stretched" with notches. Turn the taper 90° to notch the "side." which will again curl the rod. Finish notching the remaining side to once again straighten the curl. The final shaping will be done on the flat surface of the face by direct blows of the hammer.

Although the beak can help you taper more quickly, you will want to use the face of the anvil when creating slight tapering. Again, beginning at the end, you will first hammer the rod to spread its length (Picture 11).

Techniques for the Beginner Smith — Drawing Out

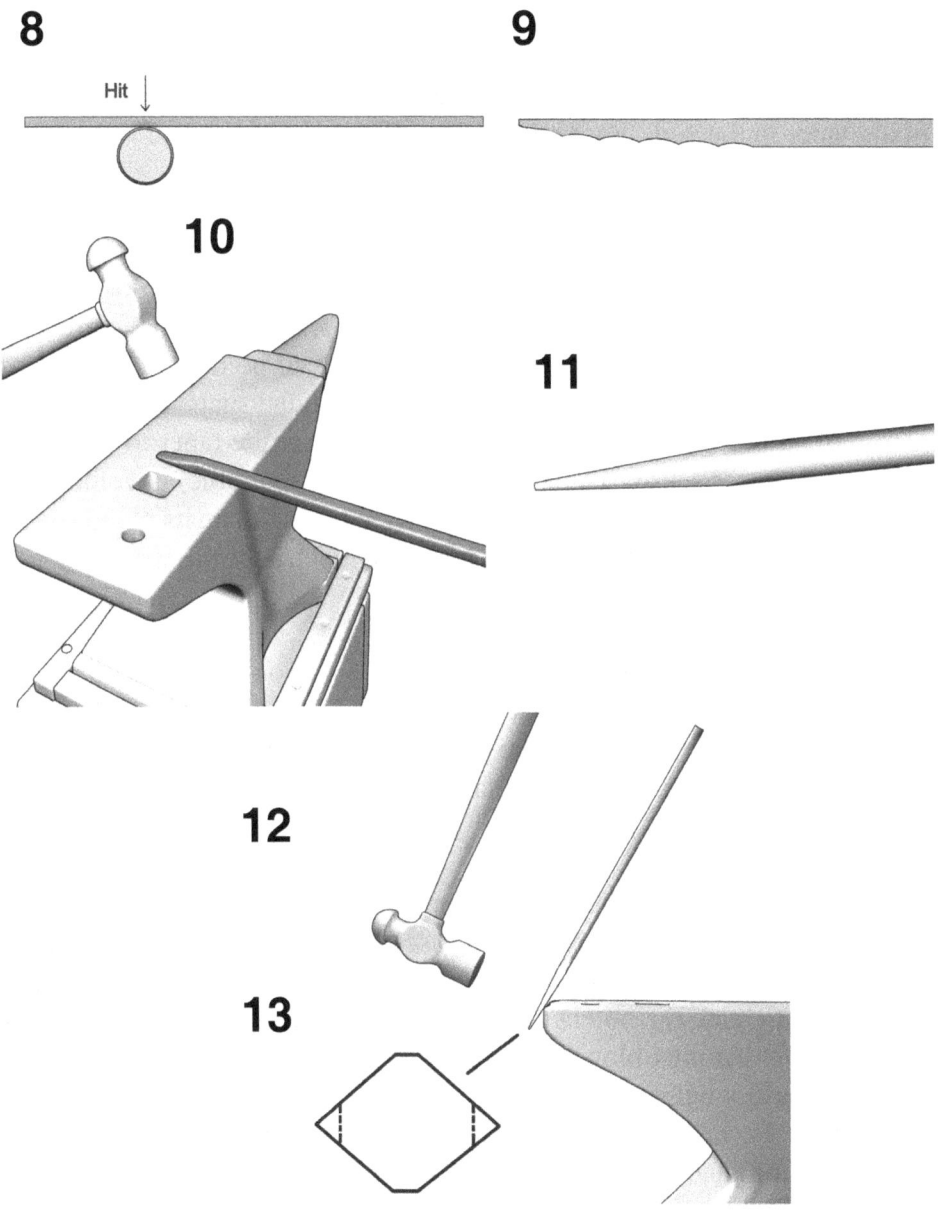

Drawing Out Techniques

You will want to rotate the metal 180° between hits to evenly reduce the thickness and lengthen the rod. If you are using a round rod, it is actually easier to work the metal when it is squared, even if you will end up with a round taper in the end. So, you will literally make your round rod square, taper it, and then make it round once again.

Even if the round rod is tapered to a point, you will have achieved this by working it into a square (Picture 12). The angles will then need to be hammered, which should result in an octagon-shaped taper (Picture 13). Additional hammering on the face of the anvil can be done to get rid of the angles, which will then return your rod to a round shape. It may seem idiotic to stray from the round shape that you want the taper to be, but it really does makes sense. Think how difficult (or impossible) it would be to deliver blow after blow to the metal while trying to maintain its round shape. Working on the metal squared gives you more control of the spreading and shaping process.

If you need to draw out large sections of a rod, you may want to use a fuller or you might just be hammering all day. The effect you will achieve is similar to using the beak or the edge of the face to create notches, but the fuller will make much bigger marks. This will enable you to complete larger jobs much more quickly.

The fuller is also an excellent tool if you wish to reduce the thickness of the rod's end without increasing its width. Use the fuller to hit the end of the rod (Picture 14) and continue to create notches along the desired length (Picture 15). As you create the notches, you will intermittently turn the rod on its side to deliver hammer blows that will bring it back to the desired width. If you don't have a helper in your shop, you will need to use a bottom fuller to complete this task because the hardie hole will hold the tool and serve as an extra set of hands. You will need both of your hands to hold the metal and swing the hammer. If the metal you are thinning is particularly thick, or the area is large, you may find it useful to use the top and bottom fullers together (Picture 16). These two tools will work together to make short work of a large task.

Techniques for the Beginner Smith Drawing Out

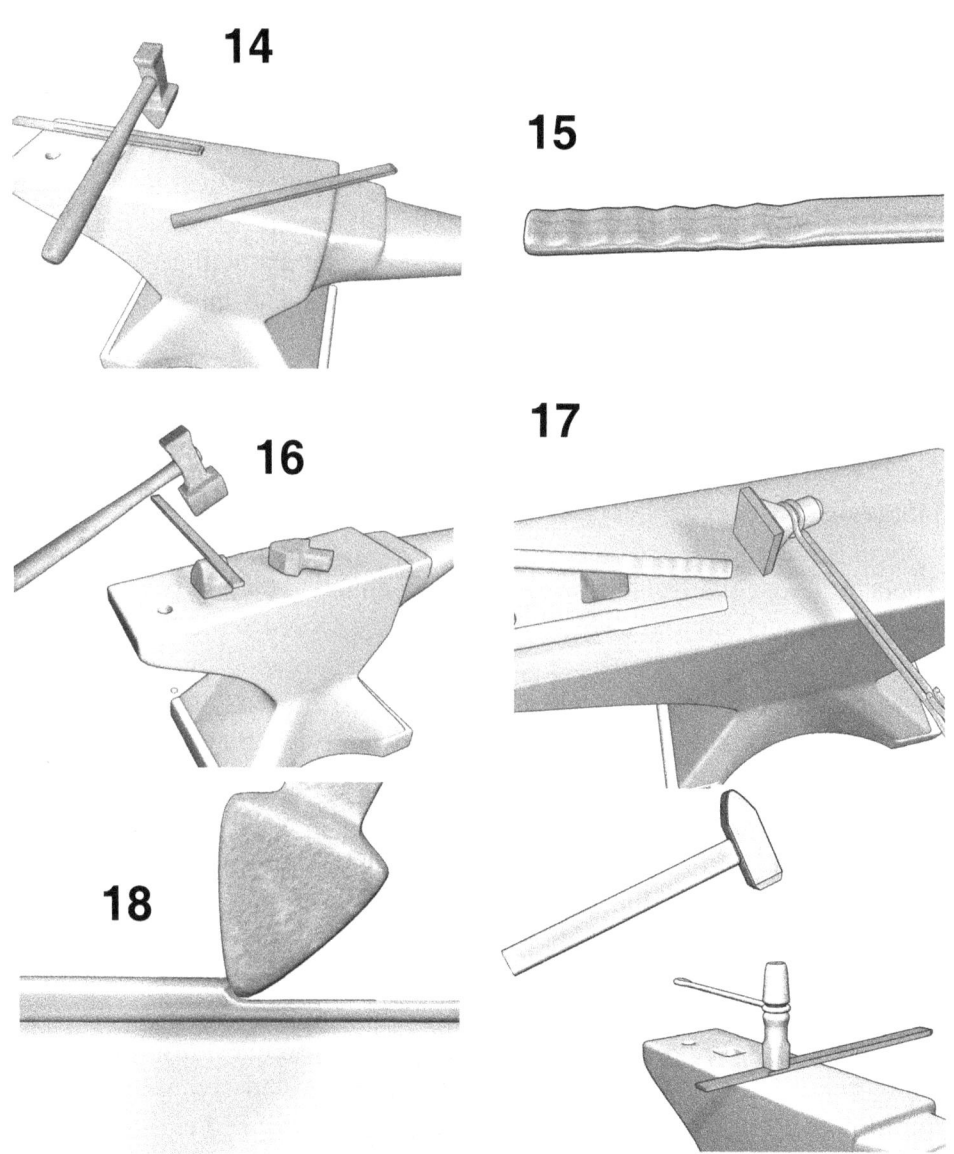

Shaping

After notching and hammering your taper you can achieve a beautiful flat finish with a flatter tool (Picture 17). You can also choose to have your flat taper blend at the point of the reduction by using a fuller (Picture 18). If you need or prefer a sharper edge, a set hammer can be driven at the reduction point to create definition. Keep in mind when finishing your piece that an abrupt change in the reduction will become the weakest part of the rod. If something heavy is lifted, it is this point that will give way first. when possible, a blended reduction makes more sense from an engineering standpoint.

Upsetting

Material and equipment: length of iron or mild steel, ball peen hammer

The act of upsetting sounds so emotionally charged, but the process is not any more traumatic to the steel than anything else you subject it to in your shop. Upsetting steel is the reverse process to drawing out, as you will make the metal shorter and thicker with this procedure. You will often find that it will be necessary to upset your steel in preparation for performing other processes. For example, the end of a steel rod may need to be upset as the first step in making a bolt or rivet. You would also need to upset the end of a metal rod if you are planning on drawing it out into a wide spread, such as you would do in making a shovel to add to your fire tools. How wide you can spread a metal rod is limited by the amount of metal there is to hammer. By first thickening the end, you will have more metal to spread.

The length of metal you are upsetting should be heated evenly to a bright, white heat. It is very important to control the heated area of the metal during this process. If you are only upsetting the very end of a metal rod, the white heat should be concentrated and not allowed to extend far from that area. If the heat travels too far up the rod, the rod will surely buckle once you start to upset the metal. If the piece you are working on is short, you will be able to dip the opposing end into the quenching bath to cool it. If the bar is too long to be dipped, you will want to have your blacksmith sprinkler handy to cool parts that cannot fit into the bath.

You can help to prevent the bar from buckling by making sure the end of the rod you are working is flat and filed. It will also help if the end of the rod is slightly beveled all the way around. This can be accomplished by filing, grinding, or hammering.

If you are upsetting a piece of metal that is a few feet long, you can imagine how impossible it would be for you to hold the end of the rod against the anvil and then hammer its end with your other hand. In this situation, you will actually utilize your own brute strength, assisted by the weight of the metal, to upset the steel rod. You will need to repeatedly raise the rod over the anvil and then bring it down with force to strike the face (Picture 19). In essence, you will "bounce" the rod off the face of the anvil, continuing until the metal is spread and shortened as desired. This particular method of upsetting steel is alternatively referred to as "jumping up", deriving its name from the bouncing action.

If you are working with a short piece of metal, you may still be able to employ the bouncing method to shorten the rod, but hammering will be more effective. The short rod can be held with tongs, or a gloved hand, and then struck with a hammer (Picture 20). Repeated heating of the metal will most likely be necessary when working with smaller pieces. Shorter rods are also more likely to bend just above the upset metal, even if you have been diligent in keeping the area cool. You will need to keep a close eye on the results of your efforts and take the time to straighten any bends on the anvil before continuing to upset the rod.

Another method of upsetting steel involves hitting the heated end with a hammer (Picture 21). After heating the end to a bright, white color, lay the rod across the anvil with the heated end hanging off the edge. Holding the cool end with your hand braced against your thigh, hit the end forcefully while turning the rod between blows to maintain even spreading. Although the view you have while employing this method allows you to observe how the upsetting is progressing, your hand and thigh will be absorbing the shock of your hammer blows. If you bruise easily, you may want to stick to one of the other methods.

Whichever method you choose, keep in mind that, in most cases, upsetting metal is best accomplished by first jumping up the rod. Once the spreading and shortening has begun, the process can be continued by hammering. This will be particularly necessary if you are trying to achieve a considerable spread on the end.

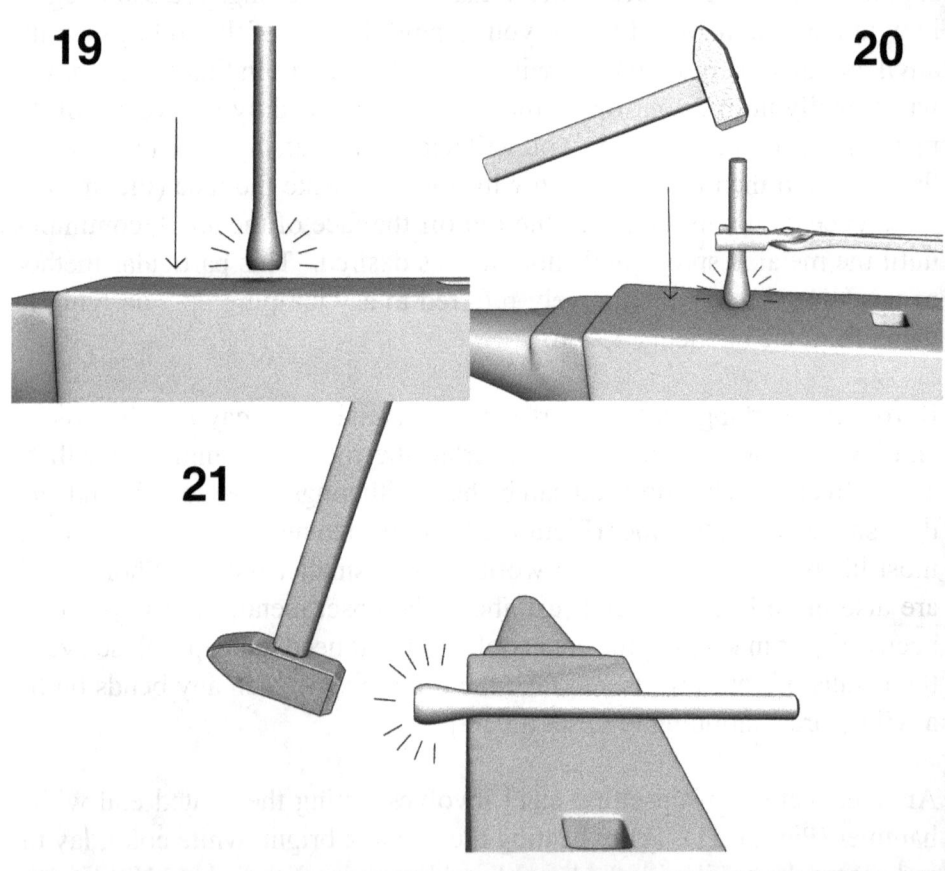

Upsetting

Twisting

Material and equipment: length of square iron or mild steel, adjustable wrench, vice, your imagination

Now, here's a really fun technique that will surely add beauty to anything you craft. You will want to use a square rod, or any other metal that has angles. Twisting of a round rod will not be noticed. If you have a round rod and want to add a twist to it, you will need to file the section to be twisted so that it is square (Picture 22). A twist can add dramatic flair to even the simplest tools (Picture 23).

Twisting metal is actually easier than it may appear. Heat just the length of metal to be twisted and place it in a vice at the point where you want the twist to end. You will then grip the point where you want the twist to begin with an adjustable wrench. It is best to twist the metal with both hands and try to maintain even movements. As you progress, you will see your beautiful twist emerge. As an alternative to using an adjustable wrench, you could use a screwing tap wrench. A tap wrench consists of a piece of metal with a square hole near its center (Picture 24). If the rod is round with a squared area for twisting, the tap wrench can be threaded onto the rod and slid down to the area to be twisted.

You can really go to town in creating twists in your creations. If you wish to put more than one twist into a long piece of metal, simply move the grips to the next area you want to twist, leaving a length of untwisted metal between them.

There are some things to keep in mind when creating twists: The key to a nice even twist lies in the heating of the metal. You must ensure that the area to be twisted is heated evenly. A "hot spot" in the rod will produce a tighter twist. Creating an even twist in a long piece of metal is more difficult than in a shorter piece. However, if the length of the metal is heated evenly your twist should come out just fine.

Most mistakes in forging can be corrected by reheating the metal and starting over, or at least making some adjustments with a hammer. However, if your twist does not come out the way you intended there is little you can do to correct it. So, in this instance, twisting is a one-shot deal. If you have other smithing to be done on a creation that will also have a twist, it is best to add the twist first. This way, if the resulting twist ruins your creation, you will not have to redo the other smithing.

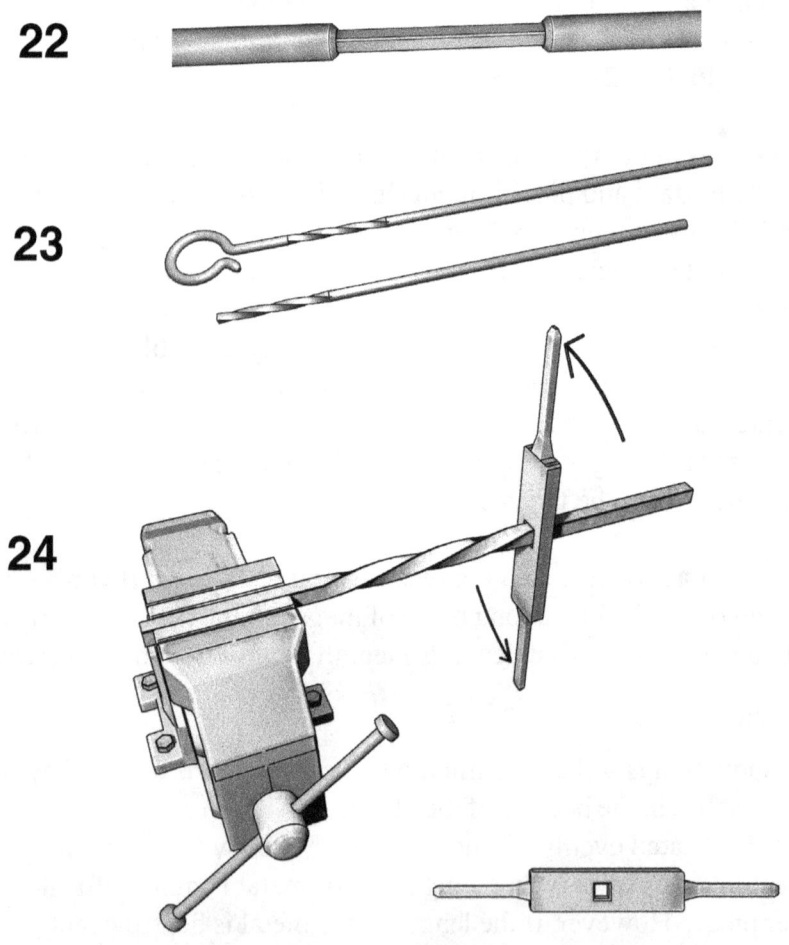

Twisting

Welding

Materials and equipment: iron or mild steel to be joined, peen hammer, a helper, flux

Blacksmith welding is uniqe and performed like no other type of welding. The metal to be joined with a smith weld is brought to near melting point and then quickly hammered together. That may sound simple enough, but considerable skill and practice is needed to perform this task well. Iron is easier to weld than mild steel, so, while you are learning, it's best to practice on iron. As you hone your technique, it will be an easier transition to mild steel and other metals. Keep in mind when choosing your practice metal that square rods weld together more readily than round ones because round rods tend to roll away from each other.

Blacksmith welding requires the use of flux to help the pieces bond together. You're wondering what the heck flux is and why it's needed? Flux is material that acts to clean the surface of the metals to be joined, which enables them to flow together for a stronger bond. Flux also combines with scale that is created by heating the metal. When a hammer blow forces the flux from between the two parts to be joined, it will take the scales with it. This process will prevent further scaling or oxidization.

Some smiths use clean fine sand as flux. This is great for welding mild steel and may be all that is needed. For other metals, however, flux made from four parts sand to one part borax is what you will want to use. In particular, when welding tool steel to itself, or to iron or mild steel, the commercial flux of sand and borax is probably your best choice. The exception to using flux would be in working with wrought iron. Wrought iron has a very high welding temperature, which melts away scale as it forms. It is possible to hammer the heated pieces together without any special treatment.

The use of flux will not necessarily create a successful weld. Too much flux is a bad thing as it can attract oxygen into the welded joint. You should only add an amount of flux that will be forced out with the first blow of your hammer. If too much flux is used, it will actually prevent the pieces from joining together. It might be useful to add a few iron shavings to the flux. The shavings will help to carry away the flux during the initial hammering. They will also burn and "collect" oxides, taking them away from the pieces being joined.

Avoiding oxides during the welding process is vital if you want your joined pieces to remain that way. Your welding fire should have a good depth burning beneath where you will place the metal to be heated and you should then cover the area with more fuel. You need to maintain a steady draft under the fire, avoiding blasts of air which would cause oxidization. Also, keep the steady draft going until the center of the fire produces a white heat that is so bright you will find it difficult to look at (sort of like staring at a really bright light bulb).

The ends of the pieces to be joined must be prepared for welding by a process referred to as *scarfing*. This means that the ends of each piece need to be upset and hammered into thick, tapered shapes (Picture 25). The surfaces should have a slightly rounded shape (Picture 26) so that when their centers meet the flux and scales will be forced out.

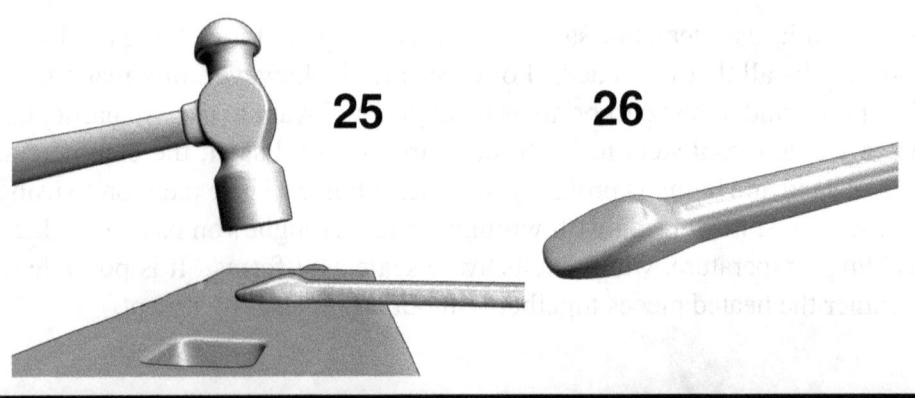

Welding

After you have prepared the pieces by scarfing, heat them in the fire until they reach an orange color. Once this color is achieved, pull the pieces from the fire and sprinkle them with flux. Return the pieces to the center of the fire where the greatest heat is produced, rounded sides up. As the heat of the fire increases, the pieces will glow a light yellow color. At this point, you will turn the rounded sides down and continue to steadily heat them. When the metal is heated to a welding temperature, a few sparks will leap up from the fire, letting you know that they are ready. At this point, the fire and the ends of the metal rods will be so bright that you'll hardly be able to look at them. Take care while you're waiting for the fire to achieve a welding heat not to allow yourself to become mesmerized and stare into the fire because, if you do, when you're ready to forge, you'll be virtually blind and it will be difficult for you to see what you are hammering! The experience would be similar to coming into an unlit building after spending time in the bright sunlight.

When your metal is sufficiently heated and you are ready to weld, you will need an assistant to hold the bottom piece of metal on the anvil. It is critical that hammering begin the instant that the metal is pulled from the fire before any heat is lost. Placing the first piece directly onto the cold anvil can be sufficient enough to cool the bottom piece and ruin the weld. Cooling can be greatly reduced by bringing the pieces together at an angle, with the bottom piece lifted slightly off the anvil (Picture 27). Your first hammer strike will force the pieces to touch the cold anvil, and also achieve the initial weld.

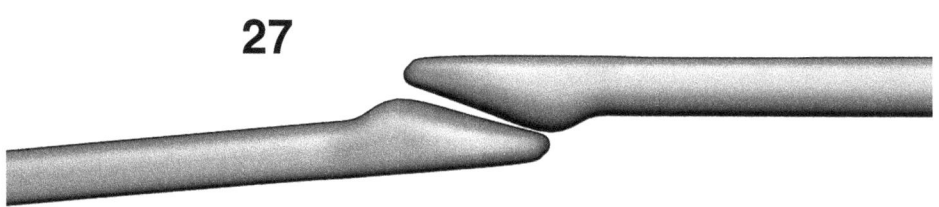

27

Welding

Hammering the pieces together must be done systematically. Your first few strikes should be at the center of the weld, hammering towards the ends of the scarf with subsequent blows (Picture 28).

28

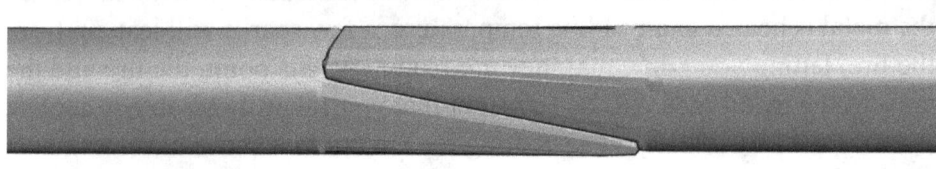

Welding

You should continue hammering around the edges of the scarf to close any openings. You will then want to turn the pieces on their sides to hammer the weld to the same width as the two joined rods. If you find that the ends did not completely close, return the joint to the fire for reheating. Be sure to remove any scale and dirt with a wire brush and sprinkle it with more flux before returning it to the fire. After reheating it to a bright white color, remove it from the fire and quickly hammer the areas that are still open.

After your weld is complete, continue to hammer the joint all over while it cools to black. Welding will enlarge the grains in steel, which can weaken the metal. Continuing with a light hammering as your weld cools will return the grains to a more normal size, thereby strengthening the bond.

This is one skill that will take practice, practice, practice! The steps in welding are very specific and speed is required to achieve a secure bond that will transform two pieces of metal into one. As you become an expert in your smithing techniques, the welding process will become second nature to you.

Heat Treatment

If you think heating and manipulating steel is cool, how about deciding how hard or soft you want the steel to be? That's right! The blacksmith has the power to make soft steel hard and hard steel soft, depending upon your needs (and perhaps your mood).

Nearly every task the smith performs involves heat, but the term *heat treatment* specifically refers to what can be accomplished by heating and cooling high-carbon steel. Heating and cooling iron and mild steel have little effect on them, but high-carbon steel can be made harder or softer as a result of how it is heated and cooled. High-carbon steel is the material of choice for making smithing tools. In fact, an alternative name for it is "tool steel."

When high-carbon steel has been forged and re-forged, internal stress points are created. The problem is not that great in iron and mild steel, but high-carbon steel is prone to stressors from forging. The stressors can be removed, however, through a process called *normalizing*. The steel is heated to a red color and then allowed to cool *very* slowly. The best way to accomplish this is to leave the steel in the dying fire and allow it to cool overnight, along with the coal and coke. It is also recommended to apply this process to mild steel that is being reworked, but the effects will not be as evident.

The process described above is also called *annealing*. This process not only rids the steel of stressors, but it will also make it as soft as possible. Softer steel is desirable if you need to file it or work it with hand tools. It's a great idea to anneal a tool that is badly worn or is slated to be forged into a new shape. For this process, the steel is heated to a cherry red and allowed to cool overnight with the fire.

The rule of thumb for these processes is that the longer it takes the steel to cool, the softer it will be.

As you may have guessed, hardening of steel is accomplished by heating the steel and then cooling it as quickly as possible. The more quickly the steel is cooled, the harder it will be. Unfortunately, this process also makes the metal brittle. Attempting to use a tool in this brittle state can cause it to crack and break, and this could ruin all of your hard work. You can reduce the brittleness (and, alas, some of the hardness) through a process called *tempering*.

The degree of tempering will be dictated by the shape of the tool and the purpose for which it was created. For a simple pointed tool, you will need to finish the end to a high shine by filing, using abrasive sandpaper, grinding and polishing, or using sandstone (a blacksmith favorite). After this is accomplished, heat the end to a full red color, but don't leave it in the fire too long. A thin tip will not take long to heat, and you should check it often for the desired color. It is best to hold it in the fire vertically with a set of tongs. Because you'll want to cool it as quickly as possible, holding it vertically will put it into the perfect position to be plunged into the quenching bath (See picture below).

When the tip has turned red, dip the tip directly into the quenching bath and stir it around to help the metal cool more rapidly. The bright, finished tip that you had worked so hard to achieve will be discolored after this process, but fret not. The preliminary polishing will help the shine return during the final finishing of the tool.

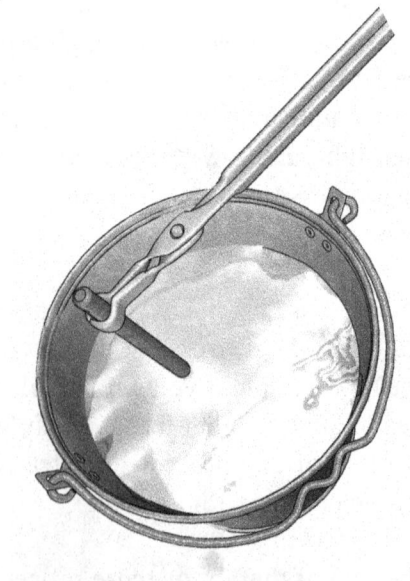

Quenching Bath

For tools that are wider than a thin tip, it is important to heat them evenly, which you can do by turning the tool in the fire as it heats. It is also imperative to rapidly cool it all over, all at once. The tool must be cooled evenly, as soon as it has reached the desired temperature, in order to sufficiently harden it.

If you don't manage to do this, the tool is at risk of cracking or becoming misshapen. Thin tools and knife blades are particularly prone to these risks. In addition, the steel may become compromised if it is left to heat too long.

Consider heating your tongs in the fire along with the tool you are tempering. If cold tongs are used, they can draw the heat from the steel, thereby creating uneven heating.

Also, consider using a propane torch to heat thinner sections of metal so as not to overheat them. However, if an overall heating is necessary, it is best to utilize an alternative approach to both the fire and propane methods. You can place an iron pan filled with sand on the fire and then place the tool on the sand. This is an excellent method to heat a thin tool evenly, although not very effective for thicker ones.

Note: Keep in mind that if you make a mistake in tempering a tool, it must be re-hardened before tempering can be attempted again.

Although quenching baths were discussed in a previous chapter, they merit mentioning again here. Consider the temperature of your quenching bath because, although cold water leads to the maximum hardness of steel, it can also induce surface cracks. Tepid water of about 60°F would probably be better even though the tool will not be as hard.

A bath of oil can also be used to quench your tools, but you must be careful not to pick an oil that will burst into flames when it meets red hot steel. Commercial quenching oils can be bought, but you probably already have the most desirable oils right in your home. Olive oil and motor oil are good choices for the quenching bath, but motor oil is much messier to work with. If you choose to use oil, consider floating its container within a container filled with water (See picture below). This should contain any spilled oil and the water can combat any flames that may crop up.

Olive Oil Use in Quenching Process

Another popular use for oil in the quenching process is to pour it directly into the water, creating an oily film on top. When the hot steel is lowered through the film of oil, it picks up the substance, taking it down into the water. It is said in blacksmith circles that this process creates the toughest steel possible, while avoiding surface cracks.

Whichever bath you choose, be sure it is full enough so that the entire tool can be submerged with room to rotate. For small hand tools, two gallons of liquid should be sufficient.

Simply stated, tempering is the process of reheating metal to a lower temperature and quenching it again. It is important to get the steel to the correct temperature with respect to the desired effect. You can judge the temperature of the steel by monitoring the colored oxides that form after it has been polished. The color will actually remain on the steel after it has been cooled, but is easily polished off.

If polished steel is heated slowly, you will be able to observe the colors as they first appear and then progress. The process begins with the color of pale straw, which will deepen into orange and then brown. The brown gives way to a reddish-brown before deepening into purple and blue. Following the blue, the metal will become red hot. Seeing just about all the colors in the rainbow makes for a beautiful experience. Keep in mind that the higher the temperature used for tempering, the softer the tool will be. The chart below shows the oxide colors you should look for when tempering your tools.

Oxide Color	Temperature F	Temperature C	Tools
Yellow	400	205	Engravers, scrapers, razors
Pale Straw	430	220	-------------------------------
	440	227	Stone drills
Straw to Orange	450	232	Saws for cutting metal
Deep Straw	470	243	Scribers, knives, punches
Brown	480	250	-------------------------------
	490	588	Dies
	500	260	Knives, plane irons
	510	263	Chisels, twist drills
Bronze	520	270	-------------------------------
Light Purple	530	275	Hammers
	540	281	Axes, center punches
Purple	550	285	Cold chisels, stone-working tools
Blue	570	300	Screwdrivers
Dark Blue	590	310	Wood saws, springs
Greenish Blue	630	332	-------------------------------

Oxide Color Chart

If you see oxide colors on a polished piece of metal, don't assume it is high-carbon. Oxide colors will also appear on polished iron and mild steel, although they will not take on the same physical characteristics as high-carbon steel when tempered.

Because much of your steel supply will be salvaged from many different places, it is sometimes difficult to know what type of metal you have. You can always employ the grinding wheel test, as described in Chapter 6, to figure out what you have. However, if the test is inconclusive, or you just want a second opinion, you can also test it by hardening it through heating it to red and then quenching. After the hardening process, try filing the metal. If the file glides over the metal without scratching the surface, you have high-carbon steel. If you are able to file it, the metal is mild steel.

Do not attempt to temper stainless steel or other specialty alloy metals. They do not behave as high-carbon does during this process, and it would just be a waste of your time.

Case Hardening

Materials and Equipment: mild steel, steel box with a lid, fireclay cement, carbonates (optional), high carbon material such as wood, charred leather, charcoal, charred bone, hooves, horns, eye of newt (optional)

Mild steel and high-carbon steel differ in their characteristics and reactions to heat because of the amount of carbon each contains. High-carbon steel contains more carbon, just as its name suggests. It stands to reason that if more carbon could be added to mild steel, it would take on the behavior of high-carbon steel. Unfortunately, there is no way for a blacksmith to add carbon throughout mild steel, but it is possible to add a layer of high-carbon steel over the mild steel by a process called case-hardening. This process is also referred to as *carburizing*. This layer is *really* thin at just 1/32 of an inch. At best, you may be able to add a layer of 1/8 inch, but don't knock yourself out.

The process adds a wear-resistant surface to mild steel and also allows for tempering. What you will end up with is a creation that benefits from both the strength of the mild steel and the hardness of a carbon-steel surface.

Case-hardening mild steel is accomplished by heating the metal while it is in contact with something of a high-carbon content. Through this process, the mild steel will actually absorb carbon, transforming its surface to high-carbon steel. The amount of heat and the length of exposure affect the degree to which the mild steel will "absorb" the carbon. With ingredients reminiscent of a witch's brew, some of the materials that are desirable to prepare the metal are charred bone, wood, charcoal, charred leather, parings of animal hooves or horns, and eye of newt. Okay, not eye of newt, but the other ingredients are legitimate. If you are a bit squeamish, commercial case-hardening agents are also available. To further promote the penetration of carbon, carbonates of barium, calcium, and sodium can be added to the mixture in small quantities.

You will need a container to place the mild steel and the carbonizing agents in. Because the materials will be heated together for a long period of time, the container must be able to withstand heat without disintegrating. A steel box with a lid should suffice or, if the piece of steel is small, an iron pipe with its ends sealed with fire clay would also work. When placing the steel into its container be sure there is ample room for each piece to be sufficiently and completely surrounded by the carburizing material.

The consistency of the carburizing agent should be in an evenly granulated form rather than a mixture of larger pieces and dust. The mixture can be a combination of ingredients. For example, charred hard wood serves well as the main ingredient. A mixture of 50% charcoal combined with equal amounts of barium, calcium, and sodium carbonates will also work well. Bone is a superior ingredient, but make sure you use charred bone as untreated bone will build pressure within the closed container.

Sealing the container is best accomplished with fireclay cement. First, place the container in the fire and heat it to at least a cherry red color. Then, pull up a chair and sit down with a good book. The container will need to be heated at this same temperature for six hours! If you heat it to a bright orange you can reduce the heating time, but even then you're looking at three hours.

Another option in case-hardening is to purchase a commercial powder mixture. One commonly available type requires that the powder be sprinkled onto the red hot mild steel, or the steel may be rolled around in the powder. The mild steel is then reheated which is what makes it absorb the carbon in the powder. This process must be repeated several times and the layer will be thinner than if the previous process was applied.

Once you have case-hardened a piece of mild steel it should then be treated as if it were high-carbon steel. After the process, allow it to cool slowly by annealing or, if it is to be hardened, quench it in water or oil. It is a common practice to reduce the brittleness of the surface through tempering as you may find it will crack under heavy use. Remember that any heat treatment you apply to case-hardened steel will only affect the surface. At its core, the mild steel will forever remain just that.

The techniques presented here are those that you will, undoubtedly, use most often. Practice them and then practice some more! In no time at all, you'll be swinging your hammer with expertise and confidence.

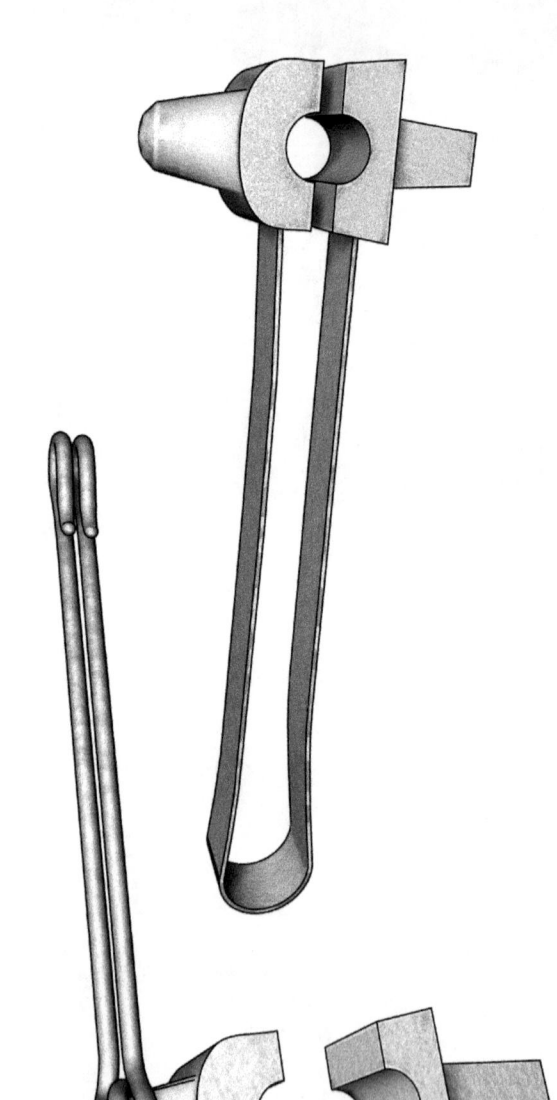

Making Your First Tools

Making Your First Tools

Well, it took eight chapters, but you're here! This is an exciting time for a new blacksmith, you can finally begin making your own tools! Of course, one of the first tools you'll want to make is a hammer so that you can toss out that soulless, store-bought one. But, don't be too hard on your purchased hammer as it served you well in learning the techniques that will now allow you to create one of your own. Once you are swinging your own hand-crafted hammer, you'll swell with pride and find that you're chest is puffed out a little further and your head is held a little higher.

It is very important that you have become proficient in the basic techniques described in the previous chapter before you begin to make even the simplest tools. Don't just try each technique one time and then start forging. Take the time to practice each one until the processes become second nature to you. While making tools or other creations, you'll want to concentrate on forming the pieces, not worrying whether or not you are performing certain tasks correctly. Fewer details will be described in making the tools, because it is assumed that you will have mastered the basic techniques.

Consider This

There are certain considerations to keep in mind when working with high-carbon steel (tool steel). Do not leave the steel heating in the fire any longer than is necessary to achieve the desired temperature. Check it often to be sure that it is not progressing past the heat color you are trying to reach. If the steel is left in the fire too long, a thin layer of the surface will become decarbonized, which will leave behind a "soft skin." The layer of decarbonized metal can be quite thin, so you might not think it will make much of a difference. However, after the steel has undergone heat treatment, the surface may not be as tough and hard as the steel below.

There is an old English adage by which blacksmiths abide when making their tools:

> If thou wilst a keen edge win,
> forge it thick and grind it thin.

What this means is that when making certain tools, it is actually desirable to have a soft surface on the steel because it will be easier to grind away. A tool with a cutting edge should always be forged thicker than you will actually want it to be. The final thickness will be determined by grinding the metal thinner. Any decarbonized steel will be removed during the final resizing. Along the same lines, when making a tool that will be finished to a point or thin edge, don't over flatten it by hammering. You will bring it down to the desired size by grinding. An extra thickness of about 1/8 inch should give you the leeway you'll need for grinding.

Decarbonizing can also occur if the steel is exposed to the air while it is being heated. You can minimize this process by heating the steel in the heart of the fire, ensuring that there are flames both below and above the metal. You will not want to hold the steel above the flames to heat it. By the same token, be careful not to heat the steel near the fire's air supply. Blasts of air meant to maintain the fire will have an adverse affect on the metal you are heating.

After you have forged your creation, you may be tempted to go right into either tempering or hardening the steel. Even if you find there is enough heat "left over" on your creation for the hardening process, it is best to separate forging from heat treatments. As mentioned in a previous chapter, the process of forging creates internal stresses within the metal. If you suddenly cool it in an attempt to harden it, there is a risk that your creation will end up cracked or distorted. It is best to normalize the steel by heating it to a cherry-red color and then allowing it to cool slowly with the dying fire. If you don't want to leave it in the fire overnight, you could also cool it slowly by submerging it into ash or sand. After leaving your new tool to cool slowly, it will be as soft as it can be, allowing for ease in non-forging processes such as drilling, filing, grinding, or polishing.

Making Your First Tools Consider This

If you want your tools to work well *and* look good, polish them with abrasive sandpaper first before polishing them to a high shine. The brightness will dim during the hardening process, but is easily brought back through subsequent polishing. If you harden steel that was not polished first putting a beautiful shine on the end product will be a monumental (almost impossible) task. Even if you do not care whether or not your tools look good, keep in mind that polished metal will allow you to see oxidizing colors during the tempering process.

If the tool you are creating is made from steel that was first formed into something else, you will want to normalize it before reworking it into a tool you will be using. Normalizing, as previously mentioned, will remove any existing tempering processes and any internal stressors left behind from the first forging. The same rule holds true for a well worn tool that you are either repairing or reworking. Sure, you could just go ahead and forge it without normalizing it, but the quality of your work will suffer. Take the extra time to ensure the integrity of your work – Normalize!

When making a new tool from a steel rod, forge at the end and leave the rest of the length to give yourself something to hold while forging. When you have completed all the work possible, you can then cut off the excess length. If you do not have enough length on the piece of steel you are working, make sure to use tongs that will firmly hold the metal. Tongs that lock on are your best bet in this situation. You will not want to fumble around with your tongs trying to get a grip on the metal and maneuvering it into a comfortable position for forging. Precious heat will be lost if you don't plan ahead.

Although you could make your own wooden handles for the tools you will craft, there are many commercial ones available that serve the blacksmith well. The handles should be predrilled with holes measuring from 1/8 to 3/16 inch in diameter and 3 inches deep. The handles should have steel ferrules (a band of metal on the drilled end), which will add strength and prevent the wood from splitting. Be sure to have a variety of sizes on hand.

The Simple Screwdriver

Material and equipment: Length of high-carbon steel - 10 inches long x 5/16 of an inch in diameter, grinder, wooden handle

The first tool presented is the screwdriver, as it is one of the easiest tools to make. In addition to its simplicity, its creation employs just about every blacksmithing technique, which makes it a great learning project for the beginner blacksmith.

The first step in creating a screwdriver involves the grinding wheel. To make the screwdriver, a piece of scrap metal will work, as would the salvaged push rod from an engine. Take the metal rod and grind the end to a point, called a *tang* (Picture 1). This point will later be driven into the wooden handle, effectively anchoring the screwdriver into the handle.

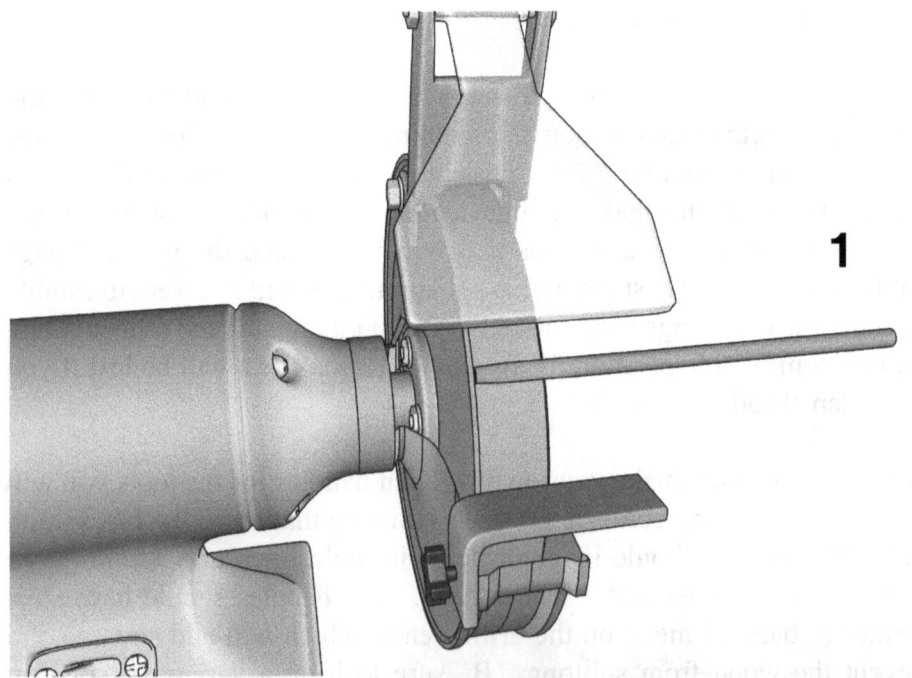

Using Grinding Wheel

Making Your First Tools The Simple Screwdriver

Use the tool rest of your grinding wheel to securely brace the metal. The tool rest should just barely touch the grinding wheel as you do this. Remember, as you move the metal back and forth to prevent grooving, keep it steady against the tool rest. At the same time, press the metal against the wheel, maintaining constant pressure. You will want to flatten four sides so that the area above the point is squared. The tang must be squared so that after it is seated in the round hole of the handle, it will be "locked" and will not turn within the handle when you use the tool.

During this process, the metal will become very hot from the friction, which is exactly what you want to happen. If it becomes too hot to handle, use vicegrip pliers to hold it in place. You want to get the metal as hot as possible from the grinding friction.

The next step is to "burn" the tang into the wooden handle. The tang should be larger in diameter than the hole in the handle. This will ensure a tight fit so that your tools don't go flying off the handle. The steel ferrule on the handle will help to prevent the wood from splitting when the tang is introduced. It is one of the few times in life when you actually want to put a "square peg into a round hole."

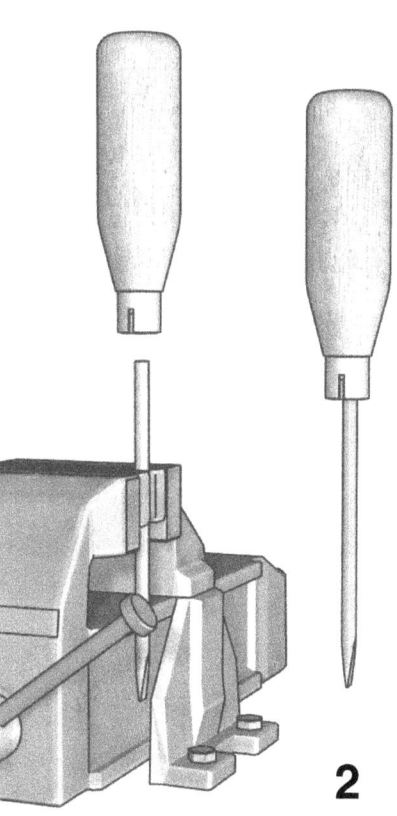

Driving Tang into Handle

Without interruption, take the heated metal and place it in a vice, tang-side up. Place the handle over the rod and tap it rapidly with a hammer until the handle has received all 3 inches of the tang (Picture 2). Depending upon how hot the tang was, you may see some smoke as it burns into the wood.

You may note that the burned wood appears soft after this process and may yield a bit, but fret not! Once it cools, it will become hard once again.

Another alternative to the squared tang is the flat tang. A flat tang might actually serve better, as there is no chance that it will turn within the tool's handle under vigorous use. Instead of tapping a square tang into a round hole, you will flatten the end and tap it into a slotted handle (Picture 3).

3

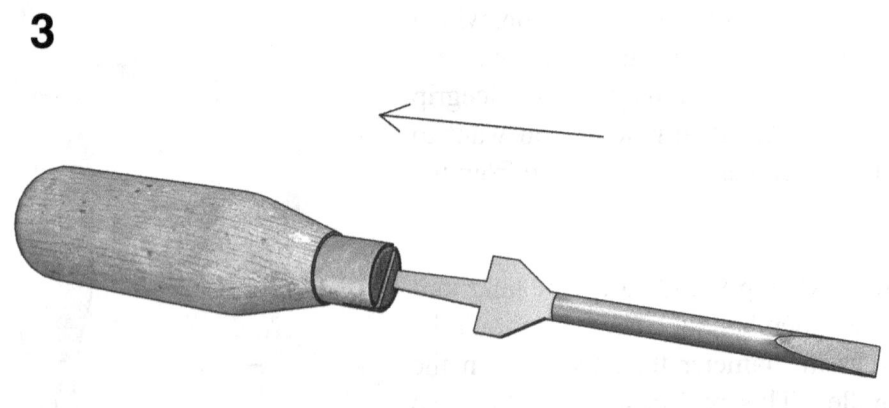

Flat Tang

You will want to make the area above the tang broader to prevent the tang from going into the handle too far. It may also be necessary to upset the steel to give you more metal to create a sufficiently broadened piece. You can also create this type of tang by first flattening the entire end and then tapering the tip.

How about a screwdriver with no tang at all? Consider making your screwdriver with a ring or oblong handle (Picture 4). Or, to achieve monster torque capability, you could weld a T-shaped handle onto your screwdriver (Picture 5).

Making Your First Tools The Simple Screwdriver

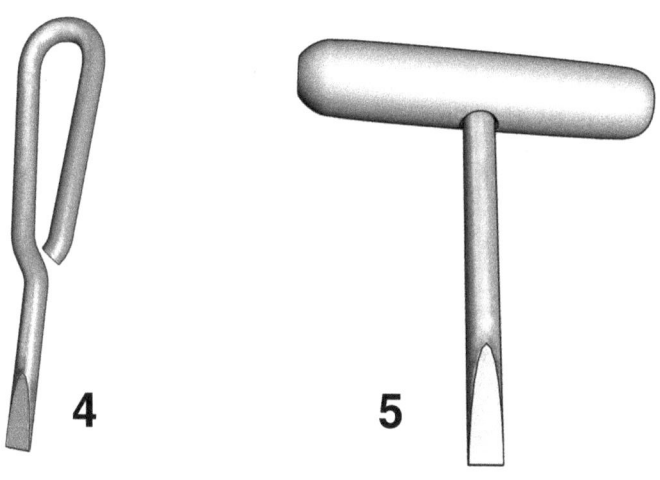

Ring and T-Shaped Handle

Whichever handle you choose, you will then need to form the other end into a flat taper by holding it against the flat side of the grinding wheel. As you move the emerging taper across the wheel you'll note that it will heat to a yellow glow. Be sure you perform this task in a semi-dark area of your shop so that you will actually be able to see the color. Continue to grind the taper until its end is as sharp as a knife.

Once you removed the tip from the heat of the spinning wheel, the color will "cool" to a dark cherry red very quickly (only 1 or 2 seconds). The moment you see the cherry-red color, quench your tool in a cooling bath.

If you find that the end of your tool chips during use, it means that the tempering process has made the metal too hard. All of your hard work is not lost, however, as this problem can be easily repaired.

Regrind the bit into a suitable taper, but don't allow the tip to heat on the grinder as you did before. Polish the end to a bright finish and then heat it to a bronze color in a gas flame (Picture 6). Then, quench the tool in a bath and that's all there is to it! The screwdriver should then be perfectly tempered and ready for use.

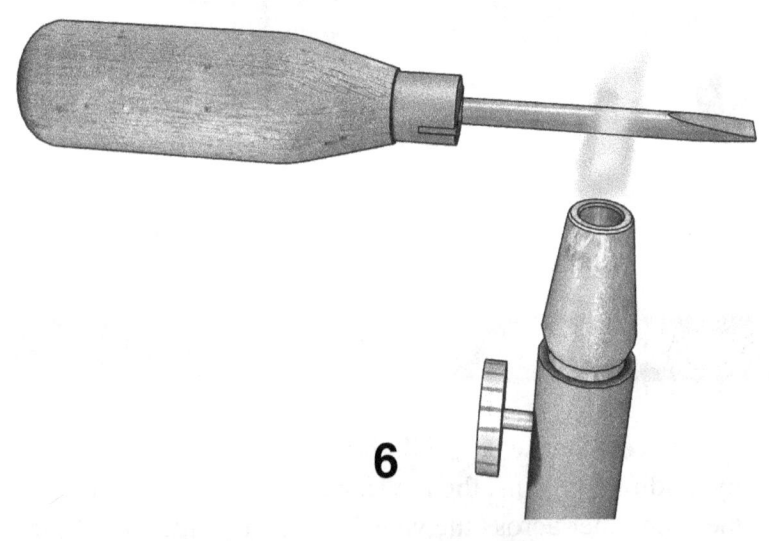

6

Using Gas Flame

The whole process of making a screwdriver should take about 15 minutes. The process for its creation can be applied to any small piece of metal that can be heated by the grinding wheel. Larger projects will need to be heated in the forge.

The Hammer . . . A Brief History

The hammer is one of the most ancient tools in existence. Dating back to antiquity, the first hammer was nothing more than a big stone that people just used to hit stuff. Unfortunately, their fingers were among the things they hit and smashed with their big stones. Somewhere around the 4th century B.C., some clever men fastened wooden handles to their rocks, using vines or strips of animal hides to hold them in place, and the hammer, as we now know it was born.

Eventually, someone came up with the bright idea of boring a hole in to the stone and found that a piece of wood could serve as a better handle. As man made his way through the ages, the head of the hammer transformed from stone to other materials available during those times. These materials evolved throughout the Bronze Age, the Iron Age, and finally, the Industrial Age that brought us steel and alloy metals.

Oddly, our colonial forefathers had quite an attachment to their hammers. After making their hammers, they often engraved the family name and date into the handle. Some craftsman even gave their hammers names, engraving "Tom" or "John" or some other special name into the handle as well. Because wooden nails were the fasteners available during this time in history, the hammers used were mostly wooden mallets. As nails became metal, hammer heads also evolved to keep up. Handles, for the most part, have always been made from wood.

Designing Your Hammer

The hammer is probably the blacksmiths most important and most frequently used tool. Designing the hammer before you begin making one is imperative if you want it to function as you intend. To design a proper hammer, you must first understand the principles of hammering.

A hammer releases energy on impact and the result of the blow depends on a few factors. For example, if hammering a nail into a piece of wood, the hammer must deliver the proper amount of energy for the task without bending the nail. The size and shape of the nail, the hardness of the wood, and the weight of the hammer must all properly fit together for the job at hand. A craftsman must have a "sense" of the ideal relationship between all of the components. Choosing the properly sized hammer can make or break a job. Is this "sense" something one is born with? Perhaps, but it can also be learned through trial and error and by learning from a more experienced craftsman. Having a good knowledge of how different shaped hammers work will also help a project progress smoothly.

Hammer Weight

The weight of a hammer must be appropriate for the task at hand. For example, if you are driving a thin nail into hard wood, several taps with a light hammer will be all that is needed. If you try to use a heavier hammer, the nail will bend like a matchstick under the weight (Picture 7). To drive a sturdier nail into hard wood, you would have to deliver well placed blows with a medium weight hammer (Picture 8). Again, you risk turning your nail into a matchstick if a hammer that is heavier than needed is used for the job. As you may have already guessed, if you are driving a heavy spike into hard wood, several heavy blows with a heavy hammer will be needed (Picture 9). If the wood you are working with is *really* hard, even the most sturdy spikes can bend. If you find this is happening, you'll need to predrill a slightly undersized hole, which will then allow you to drive in the spike (Picture 10).

Making Your First Tools Hammer Weight

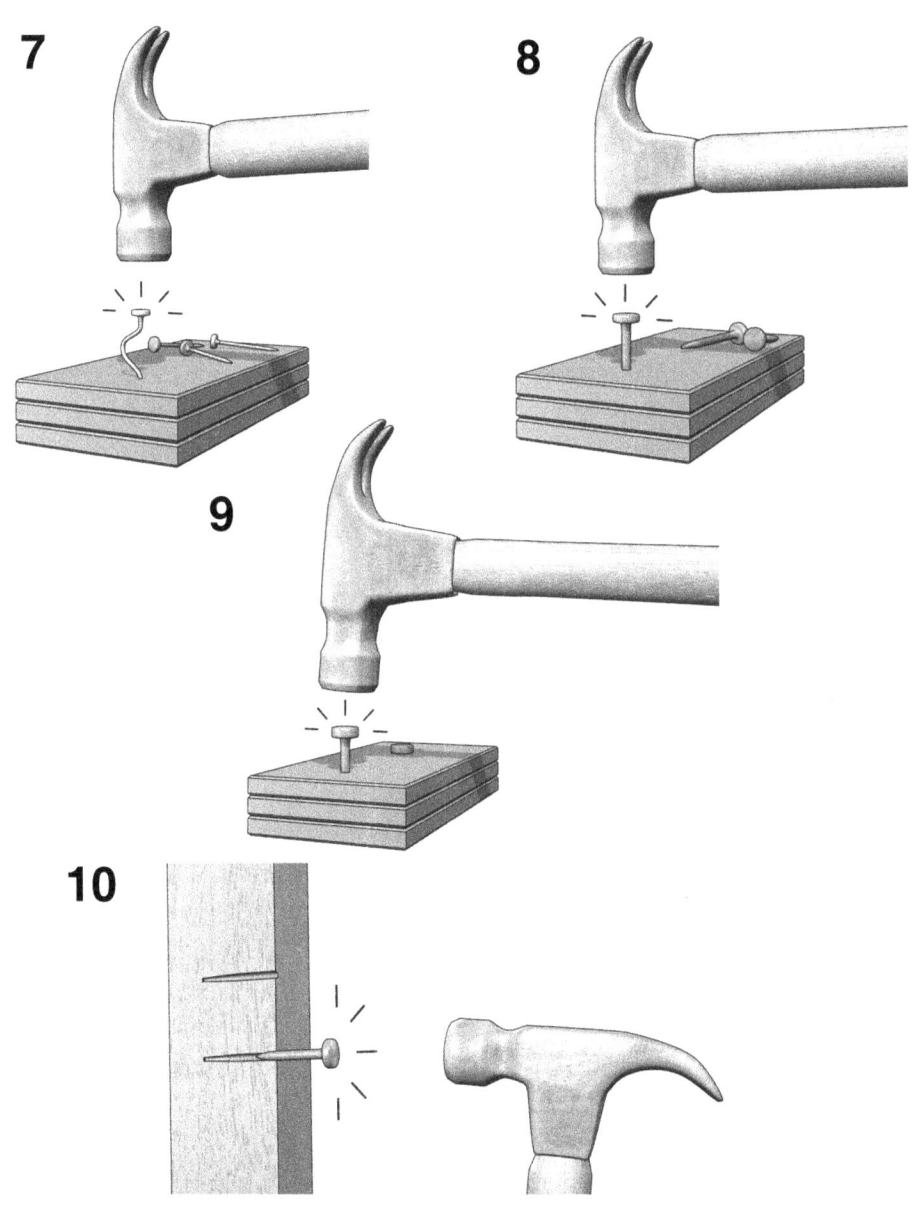

Hammer Weight

Hammer Face

The face of the hammer is the side that actually makes contact with the target. You will find that most hammers have a slightly rounded face. Now, you might think that if you have a flat-head nail, the face of the hammer should also be flat, right? Wrong!

When you bring a hammer down onto a nailhead, it is foolhardy to think that it will be perfectly centered on the head. Human beings are just not that accurate. But, a slightly rounded hammer face allows for a margin of error and will place the blow as close to the center of the nailhead as is humanly possible (Picture 11). The illustration shows a slightly rounded hammer face making "perfect" contact with a flat nailhead.

A blow with a flat hammer face would need to be exactly on the center of the nailhead, which, as previously mentioned, is impossible to control. Even a slightly off-center blow can sufficiently bend a nail or cause the hammer to slide off the nail (Picture 12).

On the flip side, a hammer face that is too round can mean big trouble. Not only will it bend the nail, it will also slip right off the nail and onto … you guessed it … your finger and thumb (Pictures 13 & 14). Ouch!

The handle of your hammer will depend upon the weight of the head. If you're making a light hammer, a short handle will serve well. If the hammer is heavy, a short handle would just wear out your arm. Basically, the heavier the hammer head, the longer the handle should be.

Making Your First Tools Hammer Face

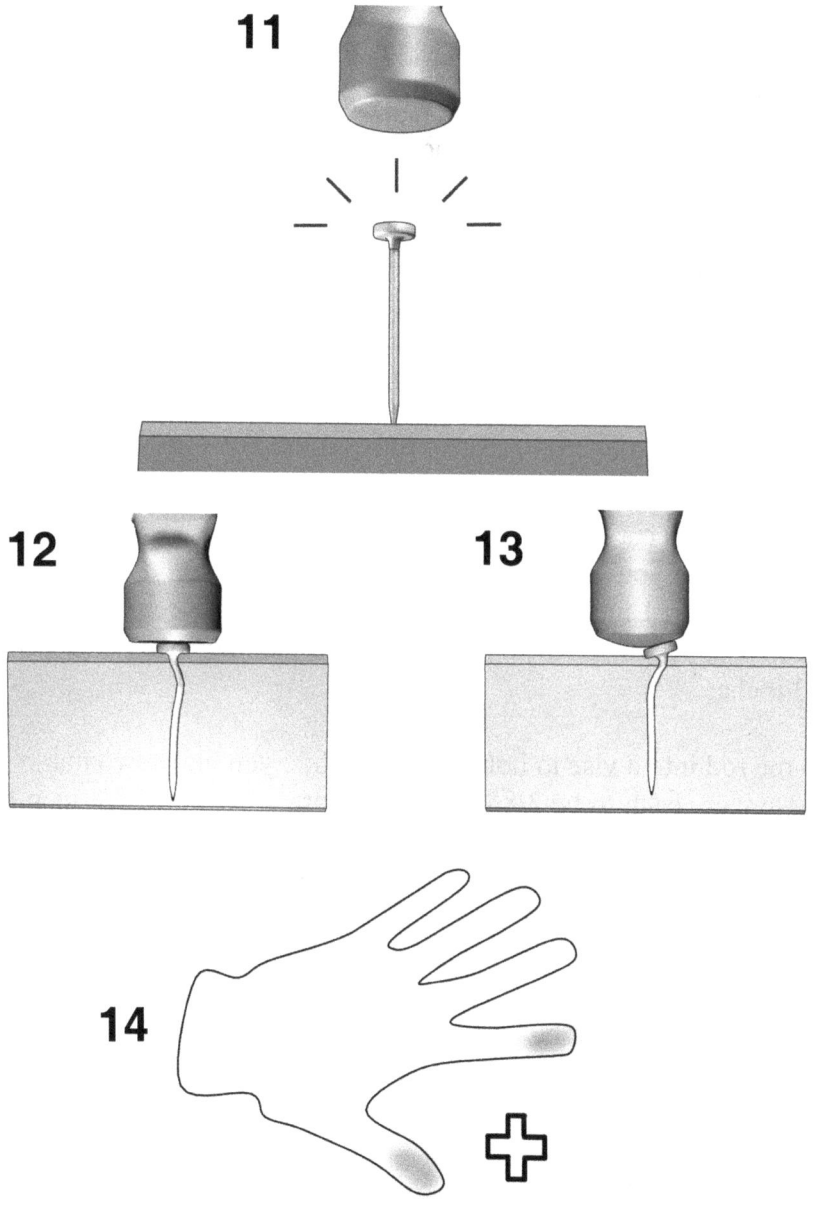

Hammer Face

Making Your Hammer

Material and equipment: Rod of high-carbon steel (3/4 to 1 inch in diameter), center punch, drill (3/8 inch bit), length of 3/8 rod, grinder, hand file, hacksaw (optional), cold chisel (3/4 inch), thick nail

The cross peen hammer is very simplistic in its design and you'll find that your shop is not complete without one. It is excellent for countless tasks, such as driving nails, bending metal, and drawing out small areas of metal. The double-duty head on the cross peen makes it more versatile than other types of hammers.

Dig through your collection of scrap metal to see if you have a square bar, a torsion bar, or a car axle. All of these will work well for crafting a cross peen hammer. If you don't have any of these objects, a rod of high-carbon steel that is 3/4 to 1 inch in diameter will do just fine. Keep in mind, when choosing your length of metal, that the final hammer head will measure about 3 inches.

Clamp the rod into a vise to hold it steady while you make two marks with a center punch, each to be 3/8 of an inch apart. Next, drill 3/8 inch holes clean through the bar where you had placed the marks. If you find the steel too hard to accommodate drilling the holes, heat it in your forge and leave it to cool with the fire. You'll recall from chapter seven that this process of annealing will soften the metal. When you mark and drill the metal, be sure that the placement of the holes ends up being exactly centered on the bar. It is important that they not be lopsided or off to one side or the other. When the bar has been drilled, the holes should end up being very close together, but should not meet (Picture 15).

Now that you have made the holes, you have to plug them back up. Take 3/8 inch diameter rods and saw off two 1 inch sections. With a few blows of your hammer, bend the ends of each one so that they will not go all the way through the holes when you pound them in (Picture 16). Once in place, grind off the excess until the plugs are flush with the rod.

Making Your First Tools Making Your Hammer

Next, you will drill a third hole in between the first two, then drive out the plugs that you just put in there. What you will end up with is a rough oval hole, which you will smooth into a perfect oval shape with a hand file (Picture 17). You will fashion the peen end of the hammer by grinding it into shape. A hacksaw could be used also, but the grinder will make shorter work of the task and allow you more control on the desired shaping (Picture 18).

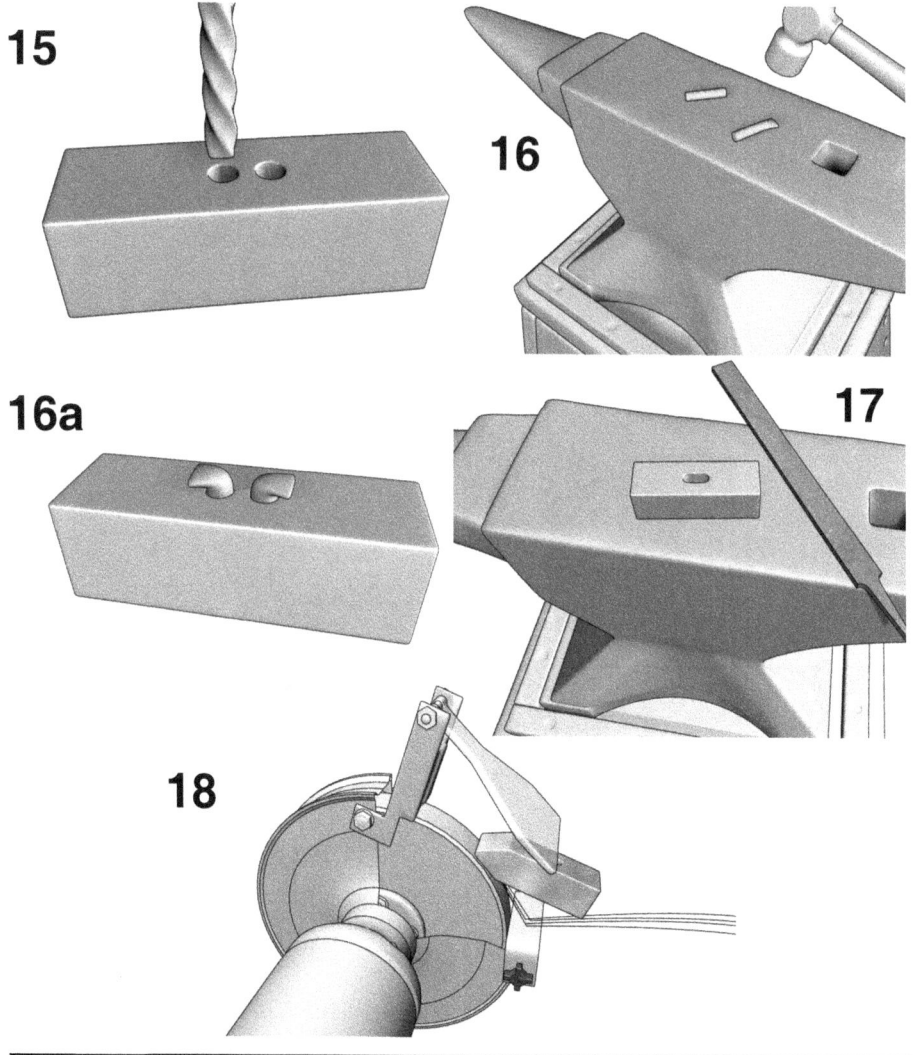

Making Your Hammer

The next step is to prepare the hole of the hammer head so that when the handle is introduced, you'll achieve a proper fit. Place the "blank" hammer head onto a sturdy wooden surface. Place a cold chisel into the oval hole that you created and, with your heaviest hammer, drive it into the hole. All that hard work you did to create a perfect oval hole, and now you have to ruin it. Well, not exactly ruin it, but by driving the cold chisel into it you should effectively splay the sides (Picture 19).

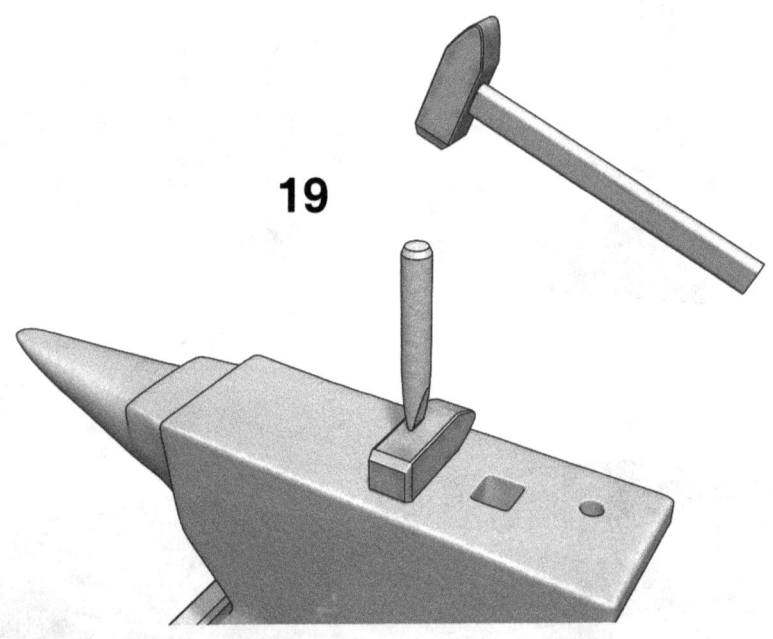

Preparing the Hole for the Hammer Head

You are now ready to temper the blank so that it will once again be hard steel. The best way to heat the blank is to string wire through the hole rather than using tongs to hold it in the fire (Picture 20). Of course, the wire will need to be able to withstand the heat. You'll find that baling wire is well suited for this task.

After you have heated the blank to a cherry-red color, quench it immediately. You may even consider stirring the quenching bath with the blank so the heat will dissipate even more quickly to achieve the maximum hardness (Picture 21).

Making Your First Tools Making Your Hammer

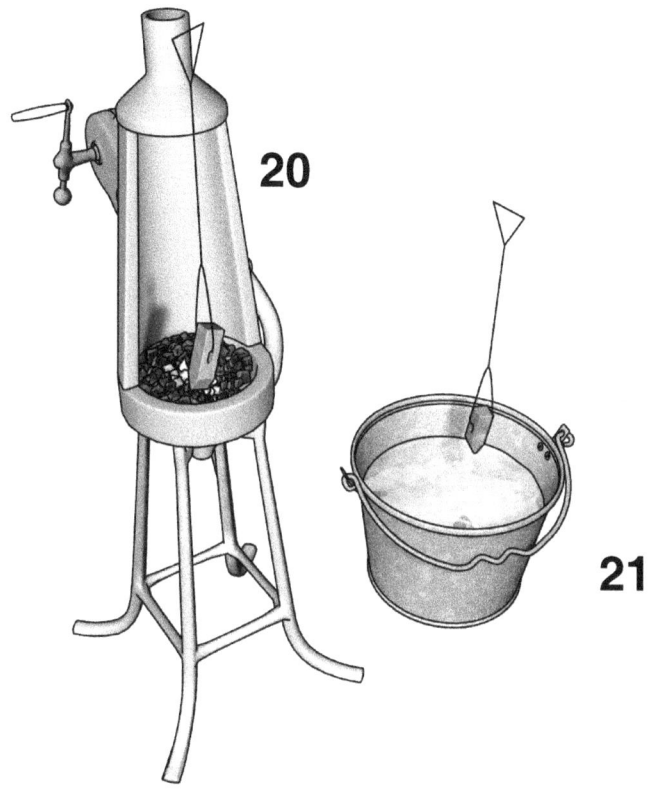

Achieving Maximum Hardness

When making your tools that will (hopefully) last for many years, consider using an oil bath to quench them. As you already know, oil will keep surface cracks to a minimum. Keep in mind, also, that by creating a hole in the steel rod, you have thinned the sides of your newly formed blank. If quenched in water, the thinner sides are at a greater risk of cracking and becoming the weak points of the tool. When using oil, be sure to have a plentiful supply (5 gallons ought to do it). You can then test the hardness of your blank with a simple "file test". Using the tip of a file, press firmly against the hardened steel. If the file slips off without making a mark, the metal is sufficiently hardened. Never drag the file along the steel. Oh, the steel will be fine, but your file will not fare the experience very well.

You will now need to anneal just the sides of the blank, which will help to prevent the hammer head from breaking during rigorous use. But how do you heat just the sides and not the rest of the blank? Glad you asked! This is accomplished while simultaneously tempering the blank as a whole.

You will first need to polish the blank to a high-mirror shine (Picture 21). Next, take a 3/4 inch diameter rod and grind its end to a taper, like a cold chisel. The tapered rod should allow you to loosely slip the blank onto it. Heat the tapered rod in the fire until it is yellow in color (Picture 22). Holding it upright, slip the blank onto it. Hold the rod with the blank on it over your quenching bath and watch for a yellow oxidation color to spread across the entire blank, from end to end. As the yellow color moves towards the ends, you'll note that the sides of the blank, which have the most contact with the heated taper, will turn purple in color. It is at this point that you will knock the blank off the rod and directly into the quenching bath (Picture 23).

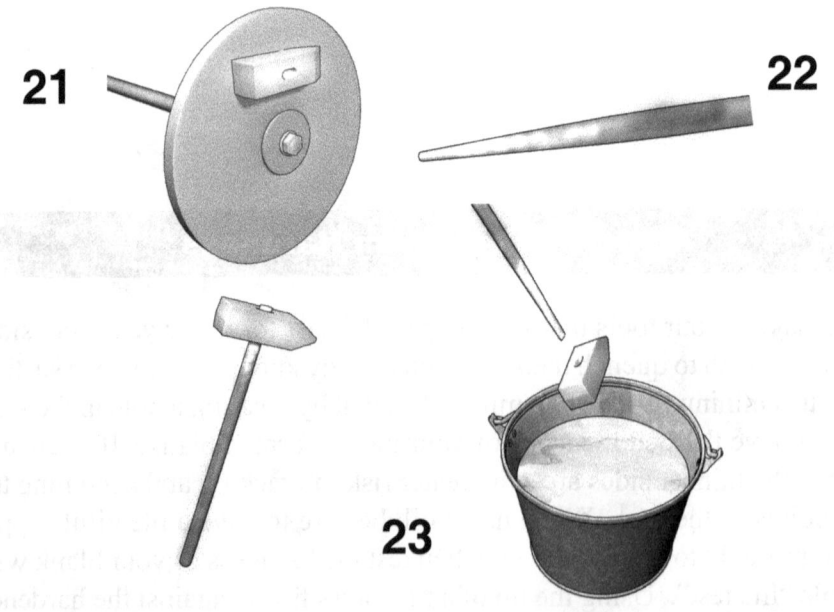

Annealing Just the Sides of the Blank

Making Your First Tools Making Your Hammer

Once the blank is cool you will need to grind the face into an acceptable surface. While grinding it on your rubber-back abrasive wheel, rotate the blank over the spinning disk, using only your wrist to perform this movement (Picture 24). This limited movement against the spinning wheel will naturally create a slightly rounded face, which is exactly what you want. You can check the shape of the overall hammer head against a square measuring tool (Picture 25).

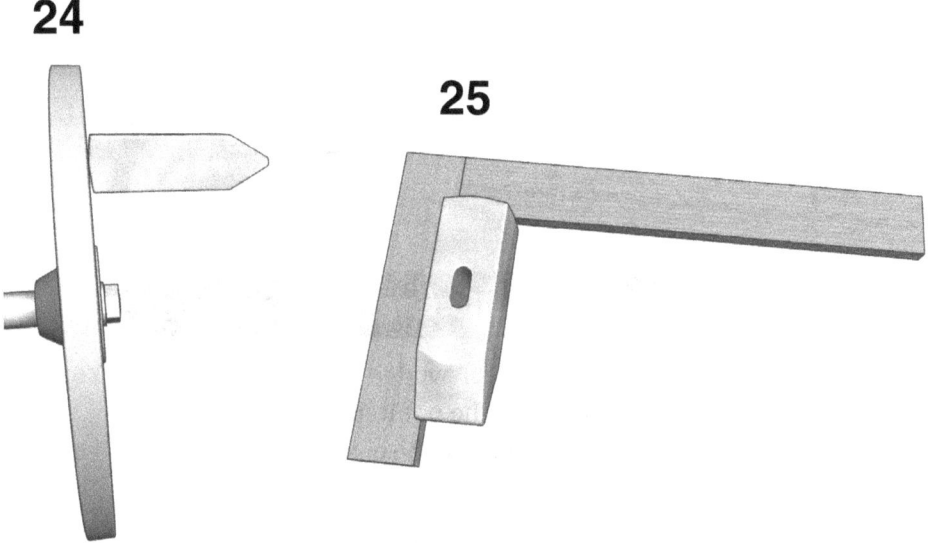

24

25

Grinding the Face of the Hammer

You have now made your first hammer head, but it won't do you much good without a handle. If you are inclined to fashion your own handle, it should be made of a hard-fiber wood, such as ash, hickory or eucalyptus. The end to be inserted into the hammer head should measure 1/2 inch wide and the opposing end should be 3/4 inch wide. The overall length should be at least 10 inches, but 12 inches is more common (Picture 26). There are also many commercially made handles available which are acceptable to use.

Don't feel like you're "copping out" if you choose to use a store-bought handle. After all, it is the head of the hammer that you poured your heart and soul into, and this is what will ultimately shape your beautiful creations. Whichever handle you choose, drive it into the smaller opening of the hammer head (opposite the side you previosuly splayed with the cold chisel).

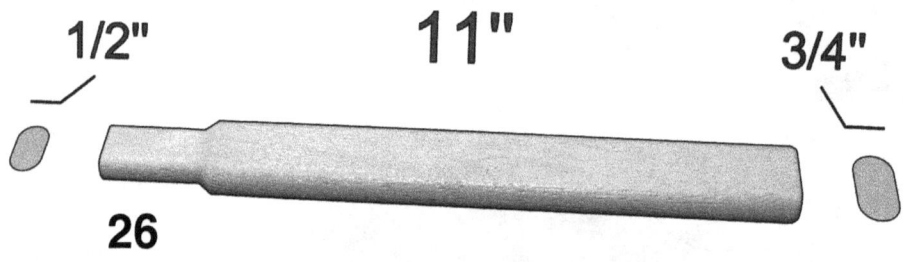

Overall Length

Next, you will need to make a steel wedge from a thick nail. You can accomplish this by cold hammering the point into a taper on the anvil. Once you have it shaped, score it with a cold chisel so that when it is driven into the wood it will hold tightly (Picture 27).

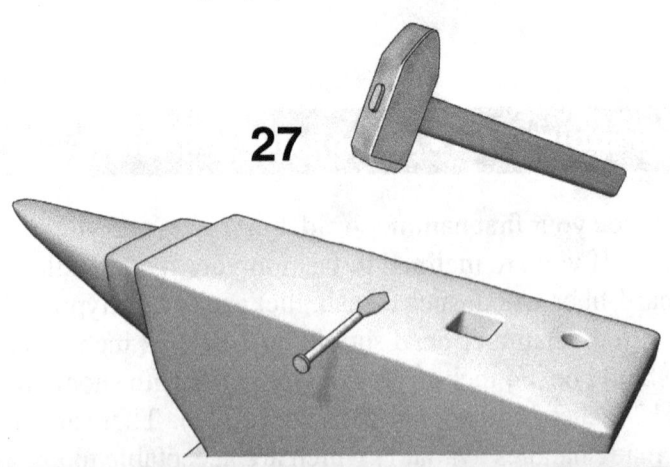

Shaping the Wedge

Making Your First Tools Making Your Hammer

Next, drive the nail wedge into the handle diagonally. This will cause the wood to spread and fill the cone-shaped hole. Be sure to hammer the wedge in as far as it will go (Picture 28). Then, cut off the excess nail with a hacksaw and grind it flush with the hammer head.

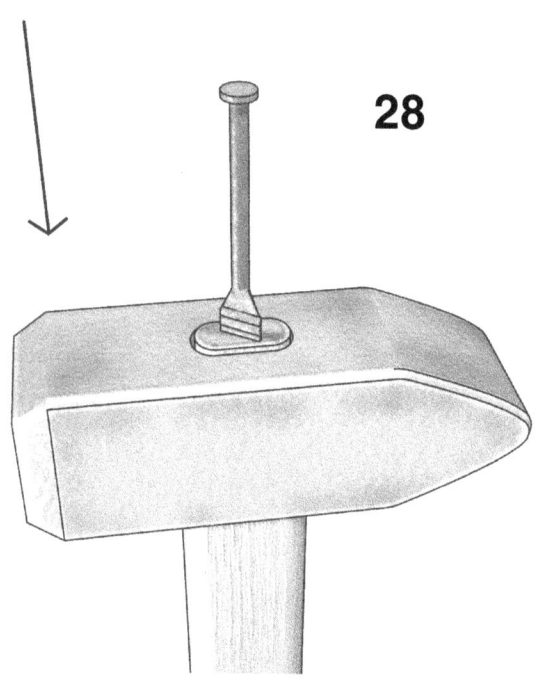

Driving the Nail Wedge into the Handle

Make sure that the handle is properly aligned with the hammer head from all angles. If the handle ended up askew (Picture 29) there is nothing you can do to refit it. You can saw off the handle at the head, punch out the wasted piece, and try using the same handle again. It won't be as long as it was before, but it may still be long enough to serve you well. If you're using commercial handles, you may just want to discard the blunder and start with a fresh handle.

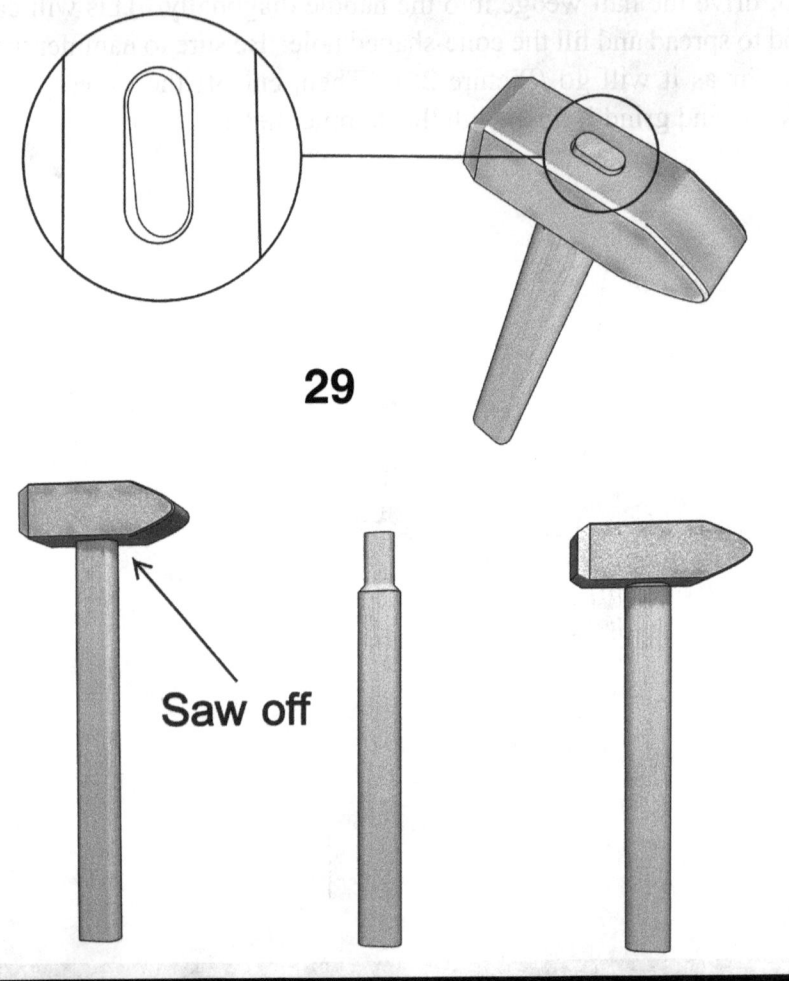

Check Handle Aligment

You have just completed the creation of your first hammer! No doubt it is well crafted and will serve you well for many years. There is nothing that compares to the feeling of pride and satisfaction you experience from having crafted your first hammer. It will forever hold a special place in your heart (and your shop). Besides the warm and fuzzy feelings you will get from your little cross peen, you have gained experience and learned the basic principles for crafting future hammers.

Bringing it Together with Heads and Rivets

Before you begin making the all-essential tongs, you need to master making heads and rivets. They anchor not only tongs, but also all hinged projects and those projects requiring bolts.

For example, if you are making a creation that has several bolts, you will want them all to be the same size. This can be accomplished with a heading tool (Picture 30).

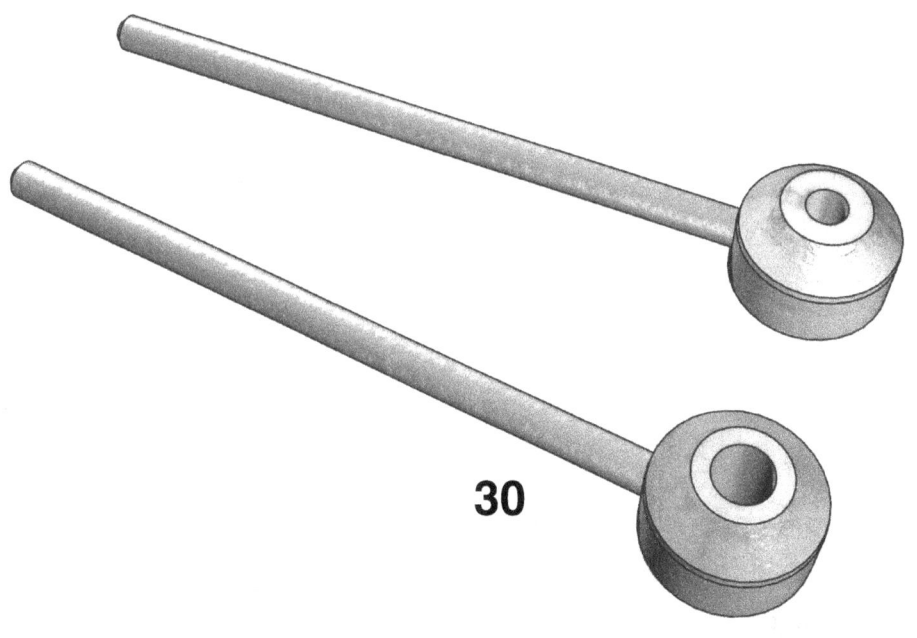

Heading Tool

A well-equipped smith shop will have several heading tools with varying sized holes to accommodate different diameter rods. The rods are heated and inserted into the hole of the heading tool and then the end is hammered into shape.

You will need a pair of rivet sets to create snap-head rivets (Picture 31). Various sets may be needed depending upon what you are crafting because different sets will make different sized rivets. The bottom set has a hollow where the first rivet head can be formed. The set can be placed into the hardie hole of your anvil, or held securely in a vise. The top set can be a punch with the same size hollow, or may simply have an opposing hollow to create the other side of a rivet.

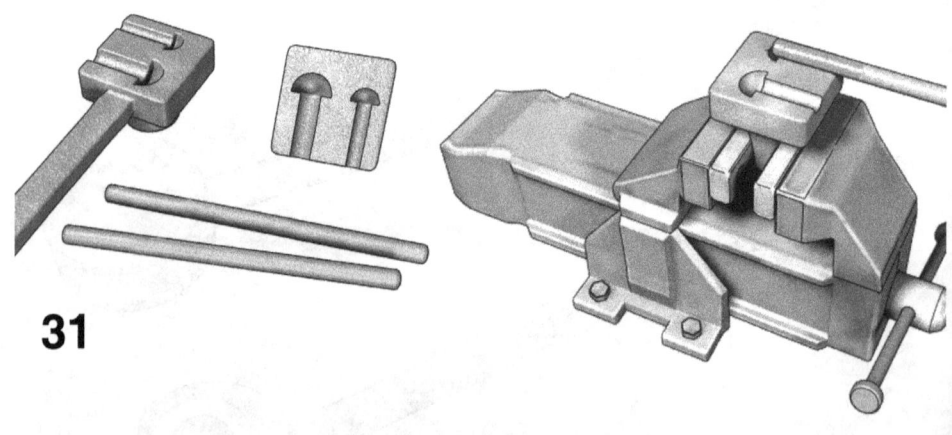

31

Rivet Sets

Swage sets come in pairs and are used to shape round or square pieces, but each set will only accommodate one size (Picture 32). For this reason, you will most likely accumulate many sizes in your shop as you hone your craft. You may be able to find a bottom swage that has several notches, thus alleviating the need to have numerous sizes. The top swage, however, will only be used for the size for which it was intended.

Making Your First Tools Versatile Tongs

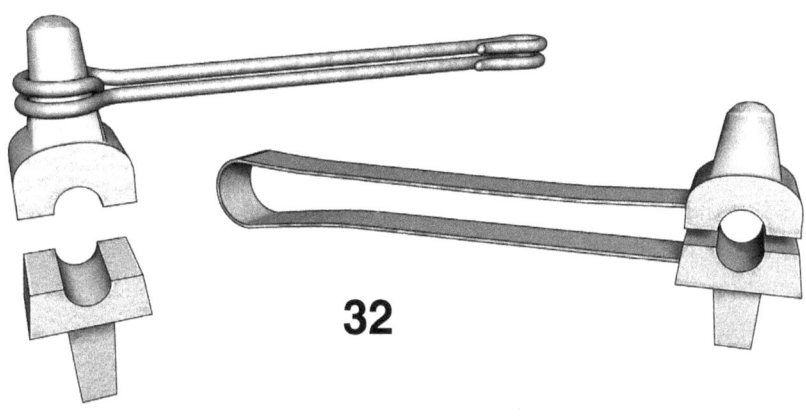

Swage

You'll note that the swage pairs can be separate or attached with a "spring." Whichever you choose, you will most likely need to have a helper. The bottom swage can be placed in the hardie hole, but the top will need to be held in one hand while you steady your creation in the other. If the work you are doing is light, you might be able to work alone, but more often than not an extra hand or two will be needed.

Versatile Tongs

Material and equipment: Length of mild steel rod (between 3/8 inch and 3/4 inch in diameter) depending upon the size of the tongs to be crafted, drill, rivet set, swage set (optional).

Although some of your initial tools will need to be purchased, tongs are tools that you can craft immediately (before or after you make your own hammer). Just as you will most likely craft several hammers of varying sizes, you'll find that you will need a wide variety of different sized tongs to suit your needs. There are some standard designs that you will want to add to your shop arsenal, but you'll also find the need for specialty designs.

You may have noticed that the material listed for making tongs specifies the use of mild steel, rather than high-carbon steel. If you think about all you have learned, this makes perfect sense. Mild steel is virtually unaffected by heating and cooling, whereas high-carbon steel is altered considerably under the same circumstances. So, while you will want the metal that is gripped in the tongs to respond to heat treatments and quenching, you will not want the constitution of the tongs to change. If your tongs are made of high-carbon steel, think about how brittle they would become from repeated heating and quenching!

Most of the tongs you'll need will generally be large. Heavy duty tongs will measure approximately 24 inches in length and the jaws will need to have an area of about 1 inch by 3/4 inch if they are to hold your creations with any stability. Light weight tongs also have a place in the shop and are a bit easier to craft. For this reason, the instruction will begin with making smaller tongs.

Consider This

When making smaller tongs, the width of the handles can remain the same as the original diameter of the rod from which they were made. You will upset the area where the jaw is formed to give you enough steel to insert a rivet and allow for proper movement. When making larger tongs, it is best that the rod be large enough to accommodate the jaw and then reduce the width of the handles.

Another method used by more experienced blacksmiths is to create the jaws and hinged area from short pieces of steel and then weld the handles on afterwards. The first method is obviously easier, but if you are making tongs that will bear a lot of weight, the welded handles will add more stability than tongs crafted from continuous rods.

As you craft your tongs, no matter what their weight or shape, apply each process to both sides before moving on to the next step. For example, if flattening a hinge area, do so for both pieces before proceeding to the next process. This will help you to make the pieces so that they will end up fitting together like a hand in a well-suited glove.

Light-Weight Tongs

The two pieces on general purpose tongs will need to be identical. Before jumping into this project, you will need a plan to ensure that the two pieces will come together and form a usable grip. If you happen to have a set of tongs that you can copy, that would be great. If you don't, it is recommended that you draw a full-sized picture of one side of the tongs on a piece of paper first. Then, trace the drawing onto another piece of paper and flip it to see if your design will be as you need it to be. To ensure that your tongs will work before you go to the trouble of making them, the jaws of your paper tongs should be closed or parallel when the handles are slightly splayed.

You should begin with a rod that is 3/8 inch in diameter and the overall length of your light-weight tongs should be about 13 inches. The jaws should comprise 4 inches (up to the hinge), and the handles should make up the remaining 9 inches.

The area where the hinge will go needs to be flattened. The small diameter of the rod will become too thin if flattened and will prove to be the weak point of the tongs. To make the metal thicker, you will need to upset the rod in the hinge area. But how do you upset a section of the rod other than the end? Well, you would first heat the area that will become the hinge. Place it in a vice at the point where you want the upset to be. When hammered, the heated section will be what "gives" (Picture 33). You will want to upset the other side of your tongs in the same way, ensuring that the thickened metal is the same distance from the end of the rod on both pieces.

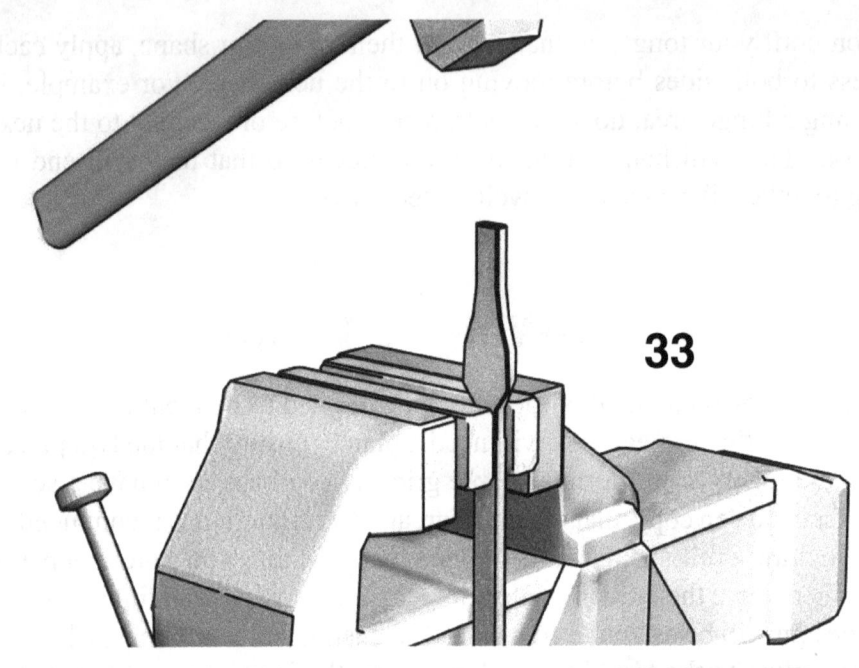

Upsetting the Rod

After upsetting the rod, you will need to create the jaws that will securely hold your creations. You should make the face of each jaw flat and make sure that they meet properly (Picture 34). You don't have to flatten all of it and may leave the back of each jaw rounded. This will save you a little bit of hammering. Make sure that the faces come together properly.

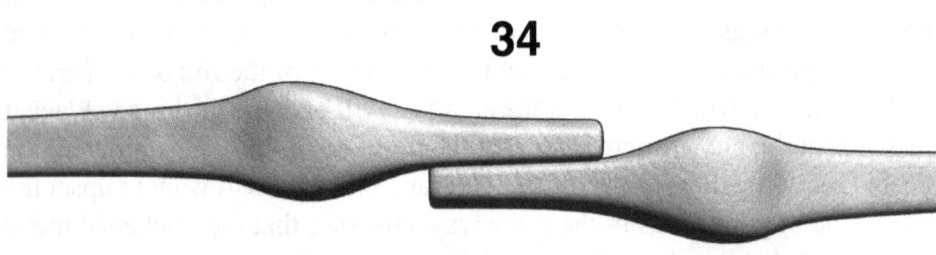

Creating the Jaw

Next, hold the piece as shown (Picture 35) and flatten the hinge area on the anvil.

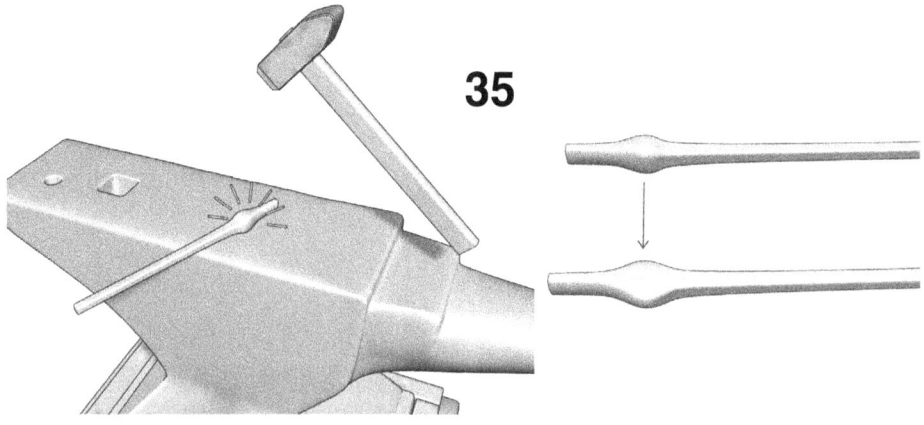

Flattening the Hinge Area

Now, on the side of the anvil, create an "S" bend so that the face and handle turn in opposite directions (Picture 36). You could leave the handles in the shape of the original rod, but curving them slightly looks nicer and will help you to keep the pieces turned properly for alignment during forging (Picture 37). This will also help you to achieve a better grip on the tongs. Check the progression of your work against your drawing to ensure that the project is coming along in the way you intend.

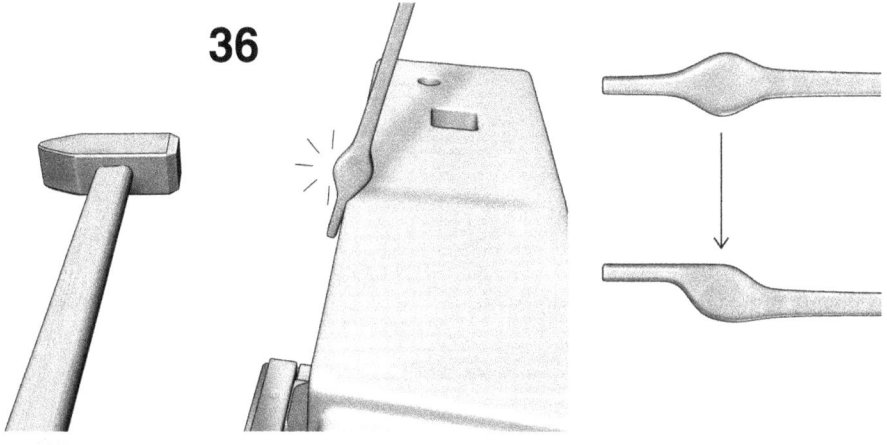

Making an "S" Bend

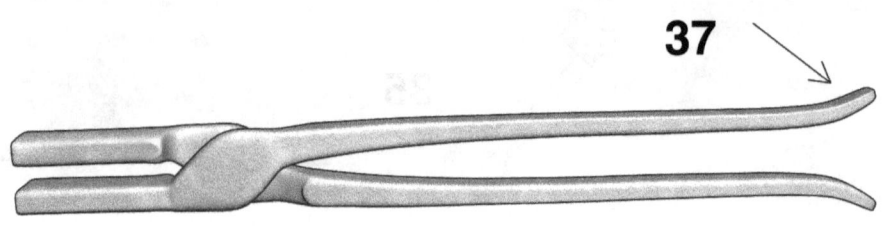

Curving the Handles

The surfaces that will meet to form the hinge must be made as flat as possible. This is best achieved by using a flatter or set hammer to smooth out the metal. Place the pieces together before riveting them. The jaws and handles should be on the same plane when the tongs are in the closed position (Picture 38).

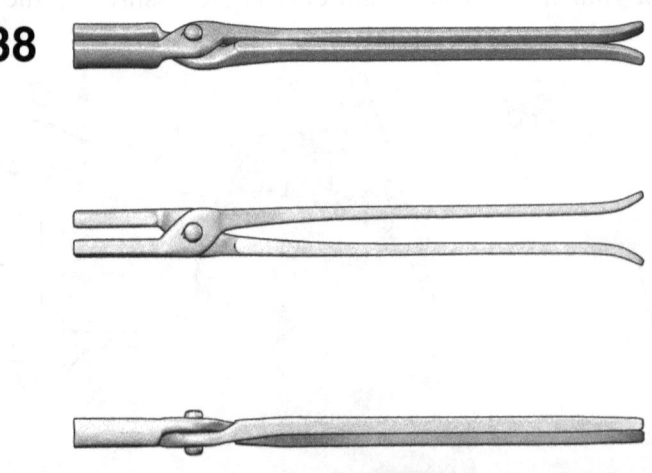

Jaws and Handles in the Same Plane

Once you are happy with the way the tongs come together, mark the hinge point with a center punch. When you drill the hole, drill both pieces at once to ensure that the holes will align. Consider testing the movement of the tongs by placing a bolt or short rod through the hole temporarily before installing a rivet.

After you install the rivet, open and close the tongs several times while there is still heat in the rivet. By moving the tongs while the rivet is still hot, the tongs will operate better at their top proficiency when everything is cooled.

Heavy Weight Tongs

To make heavy tongs, you should still begin with a drawing. It is not necessary to draw the entire tongs. A drawing of just the jaws, hinge area, and a couple of inches of the handles will do.

It is recommended that you start with square mild steel rods of about 3/4 inch thick. This will allow you to spread the jaws to a hearty 1 inch wide by 1/2 inch thick. The handles will end up with a slightly smaller diameter of about 1/2 inch.

To make the hinge, a rod of 3/4 inch should be thick enough to allow you to flatten the area without first upsetting the metal. Make sure you allow for enough length to accommodate the jaws and then hang the rod off the edge of the anvil's face to spread the hinge area (Picture 39). Once sufficiently spread, you will then need to ensure that the flattened area is "centered" along the line of the rod (Picture 40). You will accomplish this in the same way you created the "S" curve of the light weight tongs, but note the difference in the placement of the hinge area.

164 Visual Guide to **Blacksmithing**

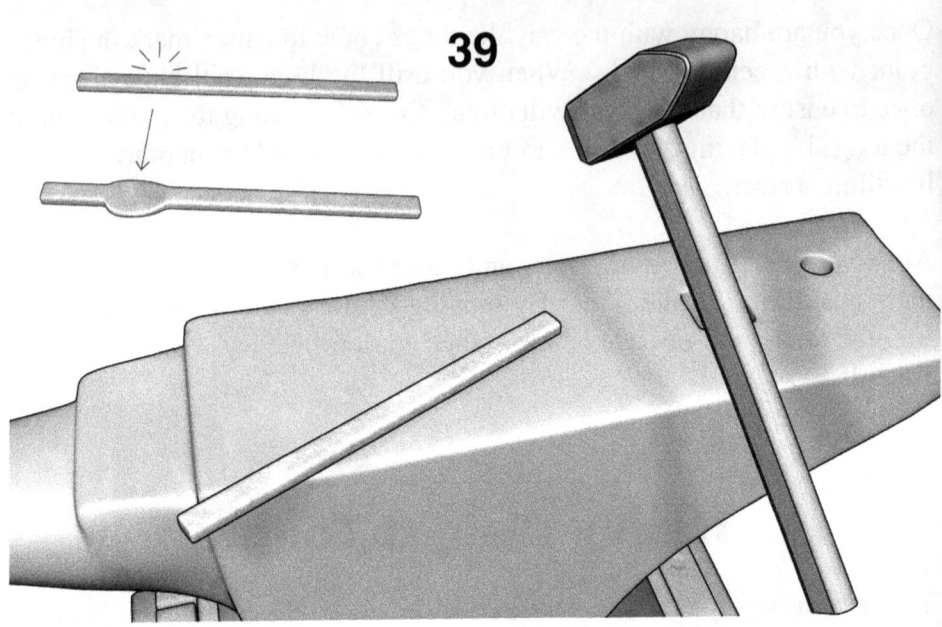

Flattening the Hinge Area

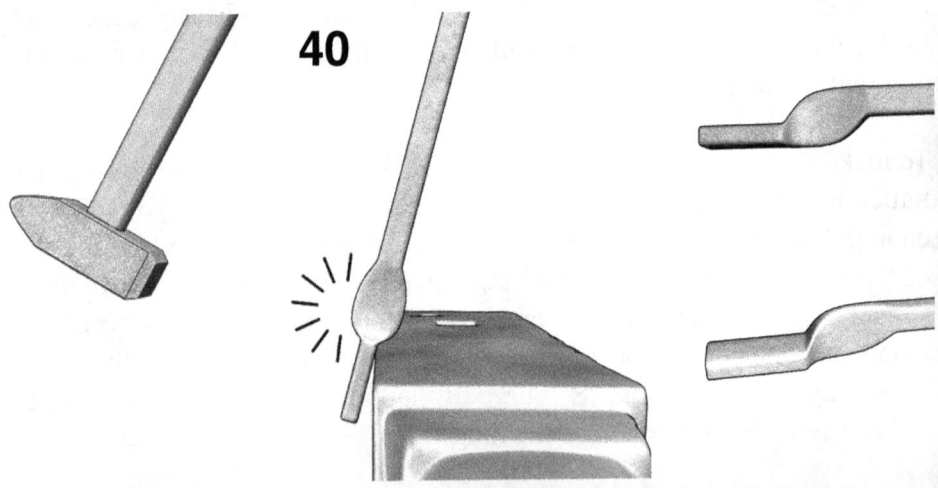

Centering the Flattened Area

Making Your First Tools Heavy-Weight Tongs

If you plan on widening the jaws of your heavy-duty tongs, this would be the next step. You can effectively widen the jaws without lengthening them much by notching the metal with a straight peen hammer or fuller (Picture 41). The notches can then be hammered out with the flat side of a peen hammer and smoothed with the use of a flatter tool.

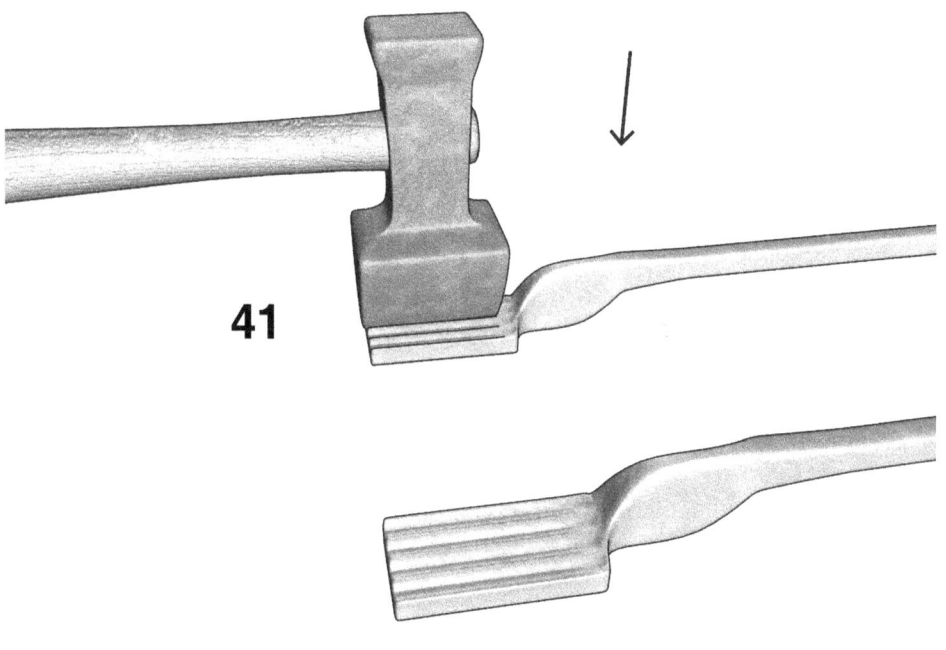

Widening the Jaws

In order for the jaws to properly meet on a pair of heavier tongs, the sides have to be offset at the point of the hinge (Picture 42). How much you need to offset them will depend upon how thick you left the hinge area. You will need to check the fit of the jaws frequently to gauge your progress as you create the offset.

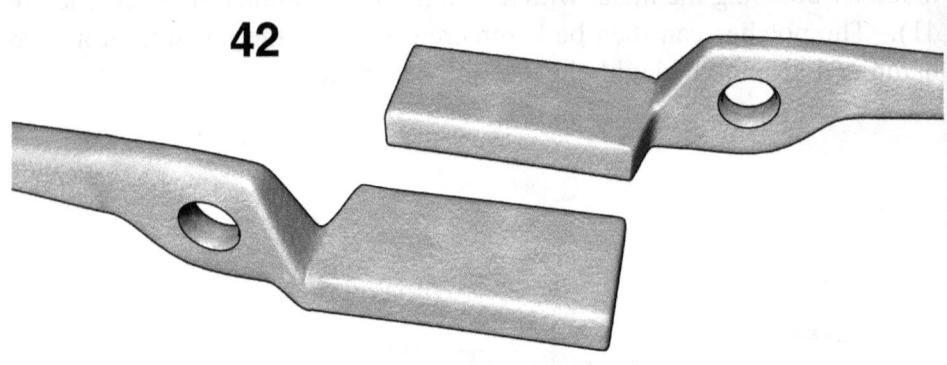

Creating the Offset

The handles will now need to be drawn out to a thickness of about 1/2 an inch. This can be accomplished through hammering and by the use of a swage set. Consider curling the ends of the handles outward for aesthetic value and ease in gripping.

If you are adding the handles after the jaws have been completed, make sure that you leave a few inches of metal beyond the hinge to give you something to weld the handles onto.

Hold the two pieces together to evaluate their fit and to ensure that all surfaces that meet do so properly. You'll also want to move the parts to make sure that the offset sections do not interfere with any other part of the tongs. For heavier tongs, it is more desirable to use a tapered punch to create the hinge hole and you will want to do this on both sides. The slightly tapered hole this creates will be filled with the expanding head of the rivet when it is introduced. The result will be a fit that is not likely to loosen through use over time. If you find this difficult, or the metal is too thick, drilling is still acceptable.

When creating your hole, it should be sized to accommodate a 3/8 inch or larger rivet. Be sure to leave a considerable amount of metal on all sides of the hole as the pivot will sustain considerable strain picking up heavy pieces of metal.

Specialty Tongs

Open-mouthed tongs with varying gaps are always good to have at the ready. When you make open-mouthed tongs, keep in mind that is it always best to have the jaws set to grip at their tip (Picture 43). If they start out completely parallel, they tend to open over time, thereby losing their grip.

43

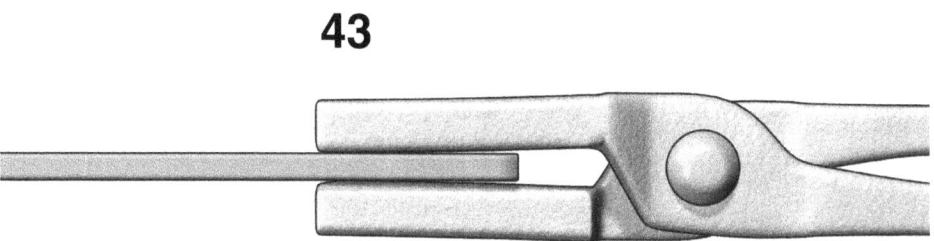

Gripping at the Tip

You may find, at times, that you need specially-shaped tongs for that one-time specialty project. So, you may assume that you'll need to take the time to create a whole new set of tongs for that one project. That's an awful lot of work and really quite unnecessary! You can always cut the rivet from an existing set of tongs, reforge and reshape the jaws, and voila! You'll have a brand "new" set of tongs.

Once you have used them for their special purpose, you can return the tongs to their original state. Keep this in mind when you create your tongs in the first place. If you leave enough bulk metal in the jaws, you will easily be able to reshape them for other purposes.

Another "trick" to getting the most mileage from your tongs is to create them with a dual purpose in mind. For example, instead of leaving one side of the jaws rounded to save some hammering, take the time to hammer them flat. You can then create a groove in the center by using a peen hammer or fuller (Picture 44). This ingenious design will allow the tongs to function as general purpose tongs, flat-jawed tongs, or tongs that will hold a round rod steady during hammering.

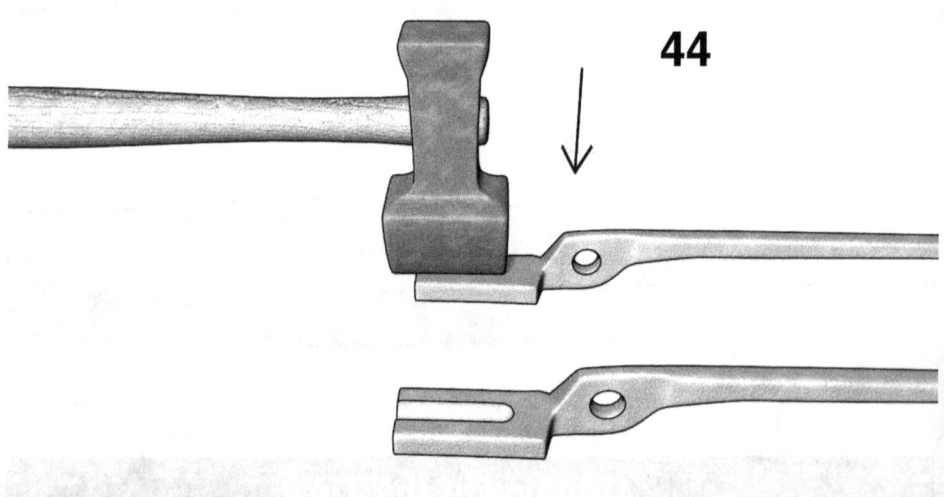

Creating the Groove

You may even find that a funky set of tongs will be useful in your shop. Offset tongs are a good example. They are designed to pick up the edges of a project and can also be used for a project that is cup-shaped (Picture 45). You would need to bend the inner jaw first and then the top jaw to meet it. There is a lot of margin for error here, as the jaws need not meet on the parallel plane. The bent section is what will be picking up the project, so this is where the jaws must meet.

Making Your First Tools Specialty Tongs

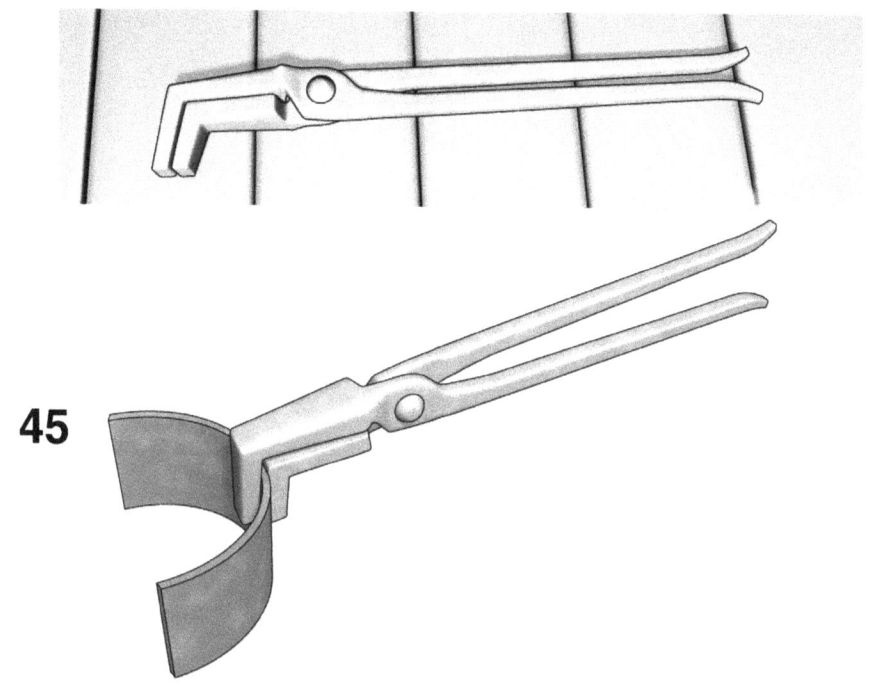

45

Using the Offset Tongs

You will likely develop a fondness for some of your tongs and they will become the ones that you use most often. It's nice to know, though, that those same tongs can be transformed to complete a special project and then be returned to their original state to faithfully continue to serve in your shop.

There are almost countless types of jaws that can be created on the end of your tongs. Each has a special purpose and you may find some really unusual shapes to suit your own needs. Coming up with creative tongs will, undoubtedly, be some of the most artistic forging you conduct in your shop.

Now that you have created some of the simpler tools for your shop, the more advanced tools won't seem so scary. Be creative when making your tools. If you need something and can't find it anywhere else, make it yourself!

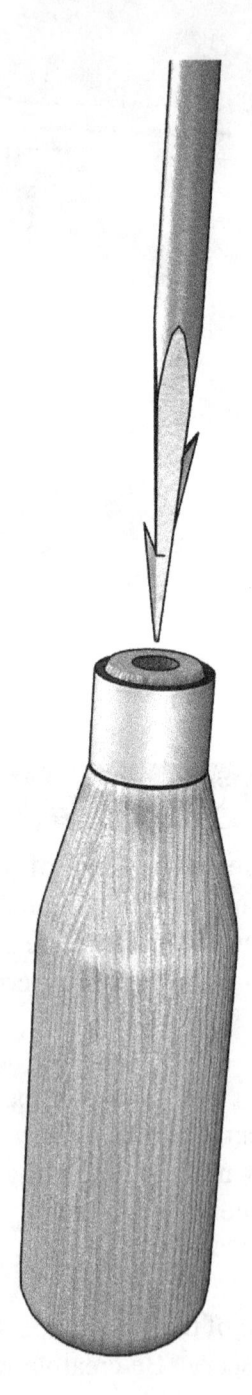

9

Getting a Handle on Things

Getting a Handle on Things

Many of the tools that you craft will need handles. Handles can be made from many different materials. You may decide to keep some store-bought handles in your shop stock, and these could adequately serve your needs. You may find, however, that your inner artist would rather make handles for the tools that you have already labored so hard to make. Excellent! Let's get started.

Consider This

When choosing the wood for your handles, it is important to use hard, fibrous wood. Some time-proven choices include black or English walnut, ash, hickory, eucalyptus, maple, and acacia. Whichever you choose, be sure that the fiber runs straight down the length of the wood, rather than across. Straight running fibers will help to transfer the energy of the hammer blows effectively, thereby saving *your* energy.

Wooden Handles

Wood is an excellent choice for handles on tools that will be heated. Because wood is a very poor conductor of heat, the handle will remain cool enough for you to grasp, no matter how hot the end of the tool becomes.

It is always wise to reinforce a wooden handle with a ferrule. A ferrule is a metal band that encircles the end of the handle to prevent the wood from splitting during drilling, tang insertion, or during heavy use. There are many pre-existing metal "bands" you can use to create your ferrule, such as an empty rifle cartridge, a CO2 cartridge, or even an empty lipstick tube.

Cut a length of band that is appropriate for the handle being crafted. Be sure to file off any rough edges that are produced during this process. Next, slightly bevel one end of the ferrule. When the bevel is properly sized, you should be able to slip it onto a slightly over-sized handle without having it cut into the wood, although it should create a "band" indentation. It should snuggly hug the handle, but not cut into it. The wood extending past the ferrule can then be cut flush and filed smooth.

If the wood you chose for your handle was not adequately dried (seasoned) it will shrink over time and cause the ferrule to loosen. If you're not sure about the seasoning of the wood, or just want added security for the ferrule, you can "anchor" it to the handle. This is accomplished by creating "dimples" in the ferrule with a center punch, effectively stabilizing it to the handle.

You will most likely begin with a square piece of wood which you will then shape into a comfortable grip handle. You can use wood cutting tools and/or file the wood smooth into the desired shape. Once you have created the perfect handle (for you), you will then fit the end with a ferrule to give stability to the wood for the upcoming processes.

Getting a Handle on Things Wooden Handles

As mentioned in the previous chapter, a wooden handle, which a tang will be seated in, should have a predrilled hole of about 3 inches. The hole must be slightly undersized in comparison to the tang. Consider creating the hole with three drill bits of varying sizes (Picture 1). The taper will help to more securely hold the tang in place.

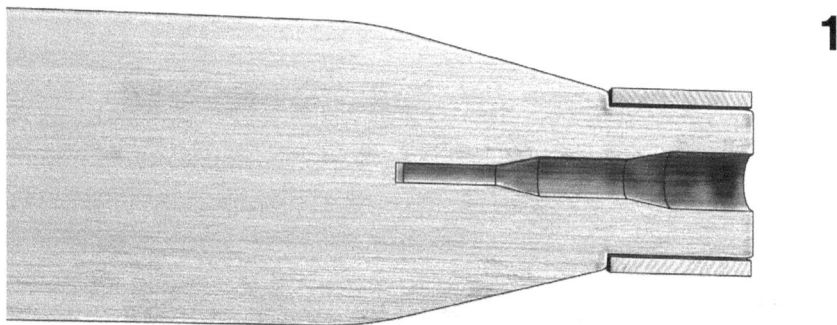

Creating the Hole

You can also add extra gripping power by making a few "hooks" in the tang (Picture 2). These can be easily created using a cold chisel and will really anchor the tang within the handle.

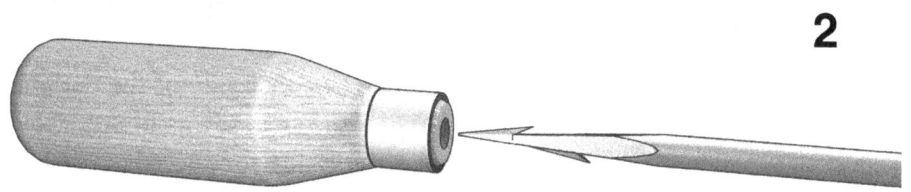

Extra Gripping Power

Another way to create a wooden handle is to apply "cheeks" to either side of the steel. In essence, you'll be creating a wooden handle, but the steel will run its entire length (many steak knives are made in this manner). You will flatten the end of the steel into a shape that is comfortable for you, although a rounded shape is typical (Picture 3). The flattened area must be large enough to accommodate two or more rivets that will secure the wood to the steel.

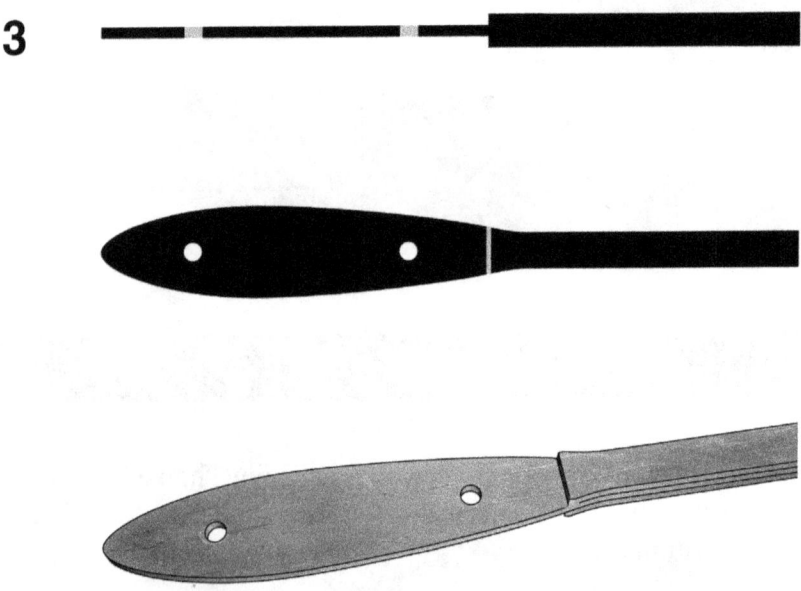

Flattening the Area for the Handles

Next, secure an oversized piece of hard wood to either side of the flattened handle (you will want the wood to extend past the steel at this point). Place the first piece below the metal and make the holes for the rivet, drilling through the steel and the bottom piece of wood at the same time. You can then turn the project over and drill through the opposing piece of wood.

Now it is time to secure the wood to the steel with rivets. You can really get creative here and add a bit of flare to your tool by using brass or copper rivets. There is no special preparation that needs to be done on the rivets.

The rivets can be hammered on alternate sides using an appropriately sized rod. This process will effectively countersink the rivets so they do not stick up on the handle and hurt your hands during use (Picture 4). Once the sides are secure, file the wood and steel so that they are flush (Picture 5). The wood can be filed into a rounded shape that will add to the comfort of the grip.

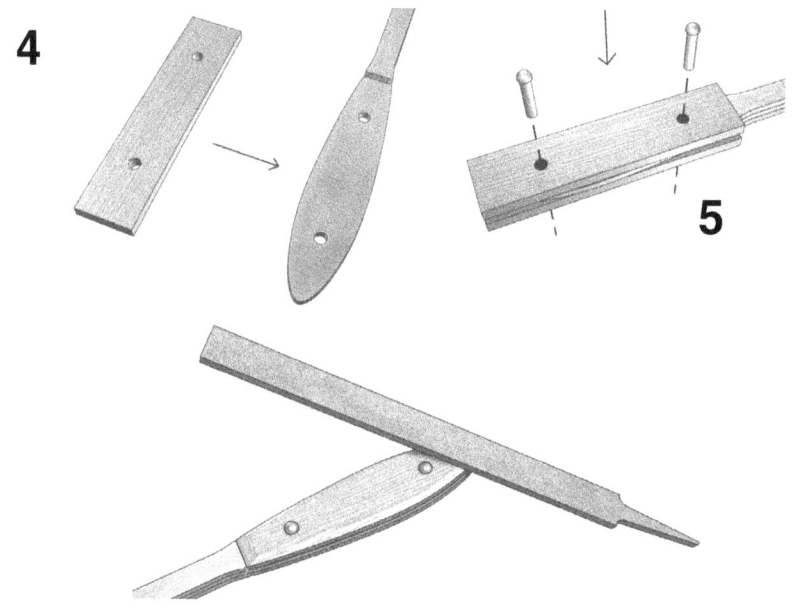

Securing the Wood with the Steel Rivets

Power Handles

More commonly referred to as "cross handles" or "T handles," these types of handles offer some serious torque. The designs of these handles are all basically the same, but can be made in a variety of ways to suit your needs and personal preferences.

A traditional cross handle begins by forging the end of the tool into a long, flattened point. The point can then be hammered into the slot of a round or square wooden handle (Picture 6).

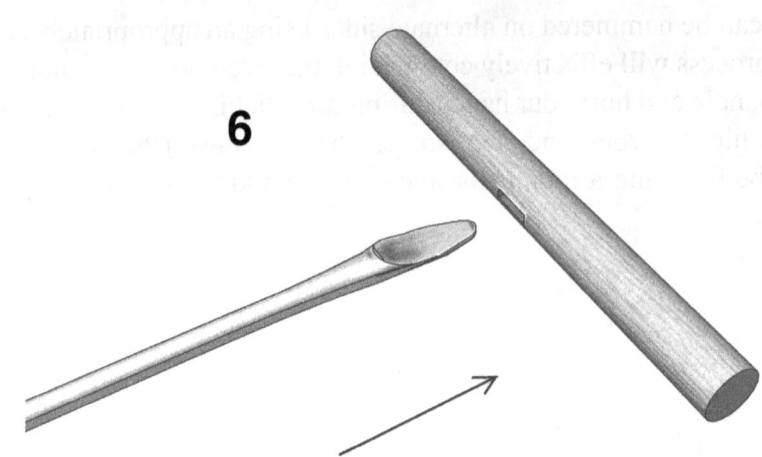

Making a Long Flattened Point

This design is quite simplistic and does not require any further processes. However, if you wish to add stability to the handle, you can drive the point through the wood completely and then use a small washer and rivet to secure it to the handle (Picture 7).

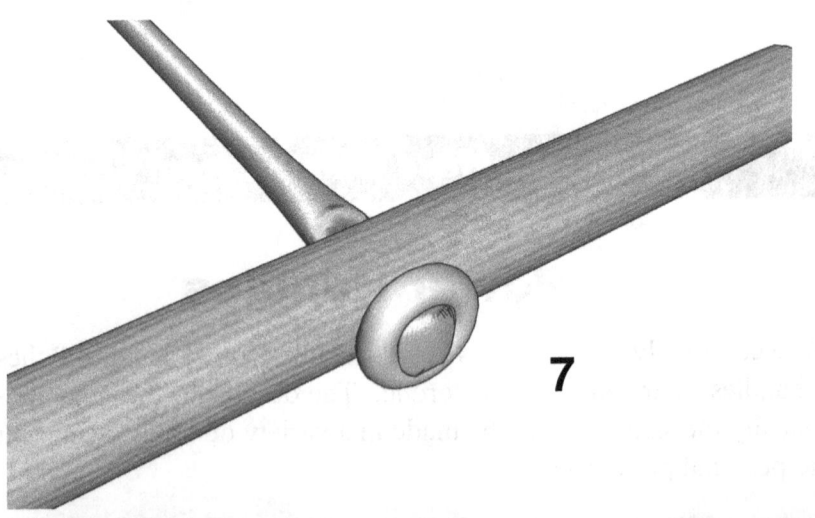

Securing it to the Wooden Handle

Getting a Handle on Things Power Handles

Beware, there is a potential problem with this design as the torque action can actually cause the wood to split. If you use a very hard wood it will probably withstand the torque, so pick your wood carefully.

Another method for adding a wooden cross handle is to create a ring that the wood can be inserted into. This design will facilitate heavy-duty twisting that is less likely to split the wood. Begin by flattening the end of your tool, allowing an ample amount to create a ring (Picture 8).

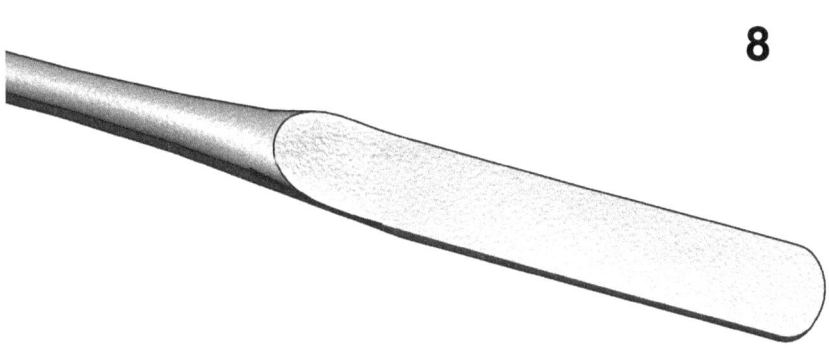

8

Flattening the End of Your Tool

Next, heat the flattened end to a welding heat. Once heated, bend it over into an oval shape (you'll work on rounding the ring after welding), sprinkle the metal with flux and weld the end onto itself (Picture 9).

Once a good weld has been made, heat the end and forge the metal on the beak of your anvil to create a round shape that will accommodate a round piece of wood. You can also use a metal rod as a handle (Picture 10).

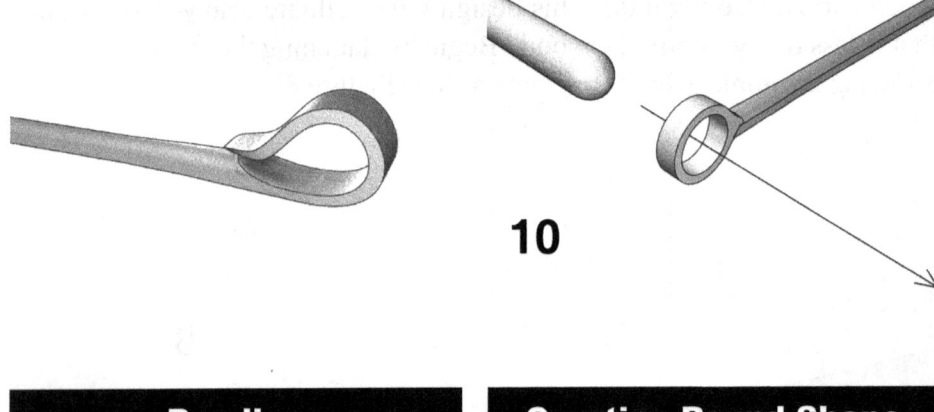

| Bending | Creating Round Shape |

For a more permanent cross handle, consider welding a steel rod directly onto your tool (Picture 11).

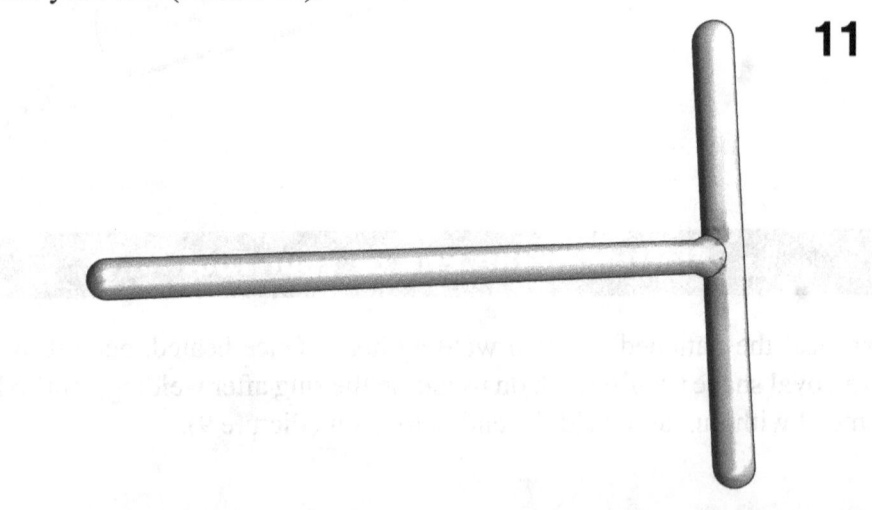

Welding Directly to the Rod

Plastic Handles

Plastic? Um ... kind of takes the romance out of crafting your own tools, but you may actually find a tool for which a plastic handle makes sense.

You'll want to shop around for a plastic that can be softened, but not ruined, by heating. The end of your tool should be left round, but with a slight taper on the end. The hole you will drill in the plastic should be only slightly smaller than the diameter of the tool.

Heat the plastic according to the directions, which is usually by boiling or in an oven. The tool will need to be heated to approximately the same temperature. Once they are both heated, you will tap the end of the tool into the plastic, but use a wooden or rubber mallet so you don't damage the plastic. Once the pieces have cooled, the plastic will have shrunk onto the metal, thus creating a nice, tight seal.

As a consummate artist, you may even discover new ways to attach handles to your tools. Perhaps you'll create a custom-grip handle that you'll make just for yourself for all of your tools. No matter where your imagination takes you, you'll gain a great sense of pride from having created your tools from end to end.

Perhaps you'll even grow to love those tools with the plastic handles.

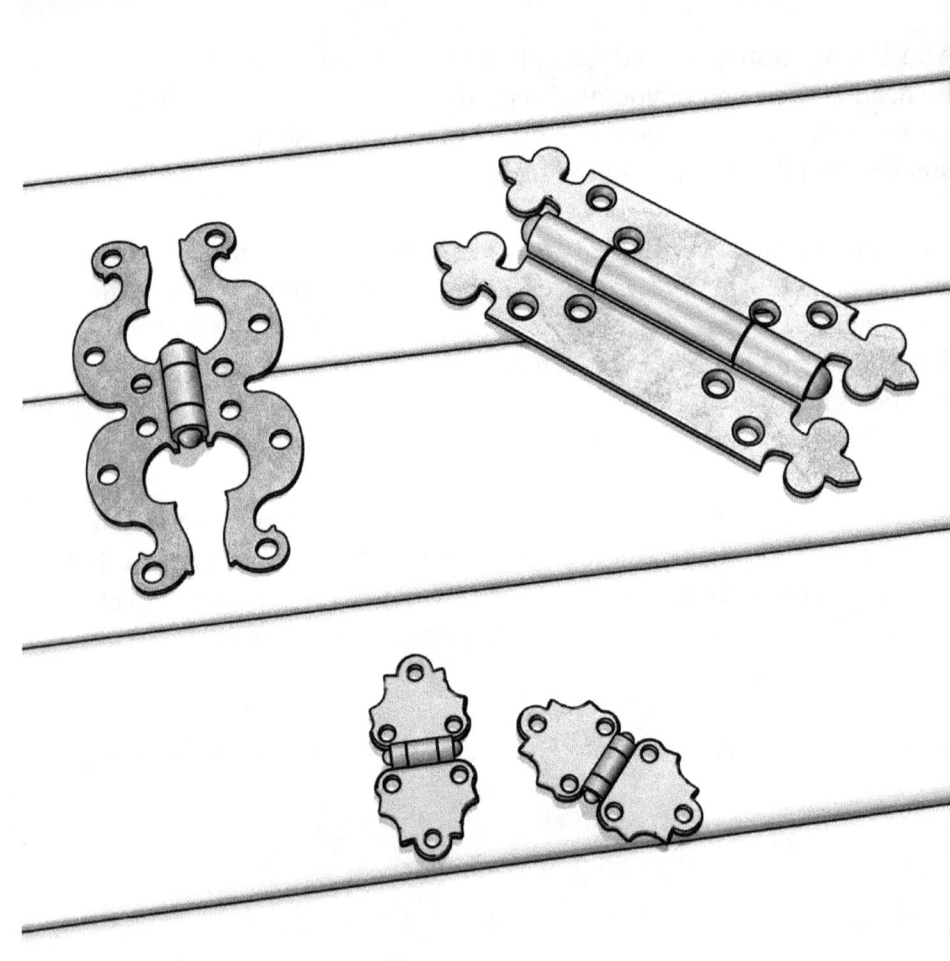

10

Making Hardware with Personality

Making Hardware with Personality

The beauty of blacksmithing lies in the ability of the smith to craft one-of-a-kind items. There was a time when a town's blacksmith worked closely with carpenters to create items that were specifically designed for their intended purpose. There were no hardware stores where replacement parts could be purchased. Although most people have fallen into the modern-day trend of buying mass produced items, there is nothing about store-bought hardware that compares to the beauty of hardware that is forged with fire and muscle power. Each hand-crafted hinge, latch, and door handle will only add to the beauty of whatever they adorn. This chapter describes how to make functional, yet beautiful, additions to your projects.

Hook and Eye

Materials: Steel rod (size depends upon the use of the hook), round or rectangular-shaped metal for backplate, rivets

One of the simplest, yet most effective, catches is the hook and eye design. They are ideal to use for holding a door, gate, cabinet, or just about anything that opens and closes. The clean design of the hook and eye makes for a great beginning hardware project. A rod will be shaped into a hook, its anchored end swinging from one eye, while the hooked end engages another eye (Picture 1). Although you may not necessarily think these primitive devices will look beautiful on your finished projects, they can be decorated with a one-of-a-kind design that is all your own.

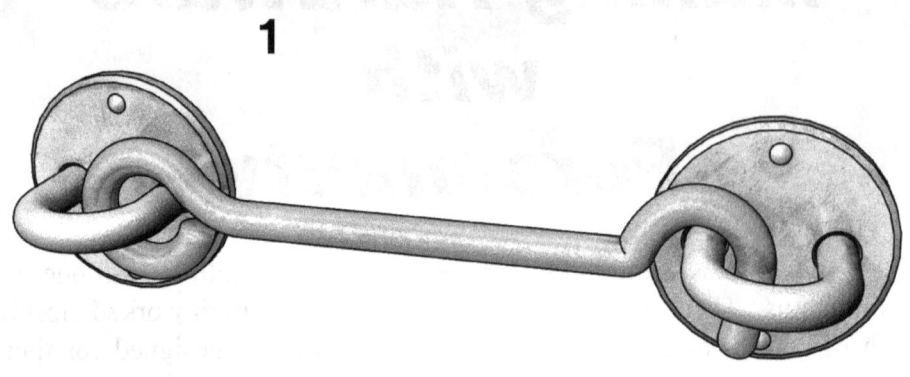

Hook and Eye

For larger catches, the rod diameter should be 1/4 or 5/16 inch. You will need to shape two rods for the eyes into a U and shoulder the ends so they can be fed through the backplate and be riveted down (Picture 2).

Shaping Rod into a "U"

Think of the eyes as big staples. They will attach to the backplate in the same manner that a staple remains fastened to pieces of paper. The easiest way to shoulder the ends is by filing. After forming the U-shaped rods, place the eye into your vise so that both ends protrude above the jaws. You should also place a washer on the end you are filing to help guide the file and ensure that the ends are symmetrical. The washer will also save your vise jaws from being marred by the file (Picture 3).

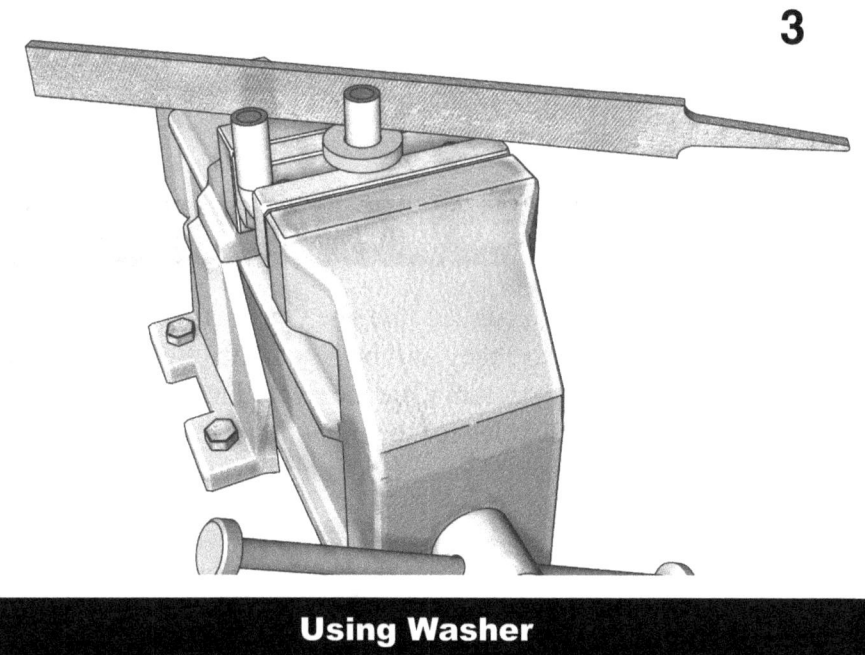

Using Washer

If you are crafting a large eye and are finding that the filing is taking forever, consider using a swag to shape the ends.

You should now prepare your backplates to accept the eyes you have just created. If you plan on decorating your backplates, now would be the time to do so. Scalloped edges dress up a backplate nicely and will show all who admire it that it was indeed hand-forged. This is accomplished by striking the edge of the plate that has been heated red hot. Keep your hammer blows uniform to achieve an attractive, even edge (Picture 4). Be creative! You can also decorate the overall plate with a bit of imagination.

4

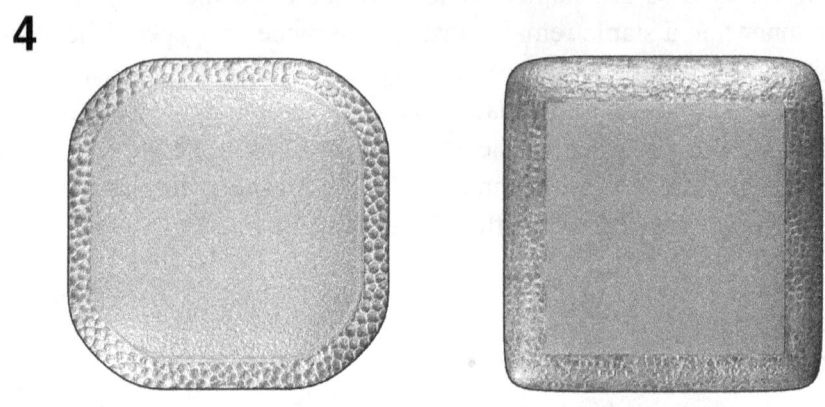

Decorate the Backplate

Next, drill two holes, using your eye (the one you created, not the ones you see with) as a guide to ensure they will be a good fit. Be sure to include a countersink on the back to accommodate the rivets that will hold your eye in place. Drill two additional holes for the mounting screws, again including a countersink, on the front of the plate (Picture 5). Your finished backplates will have 4 holes (Picture 6).

5 **6**

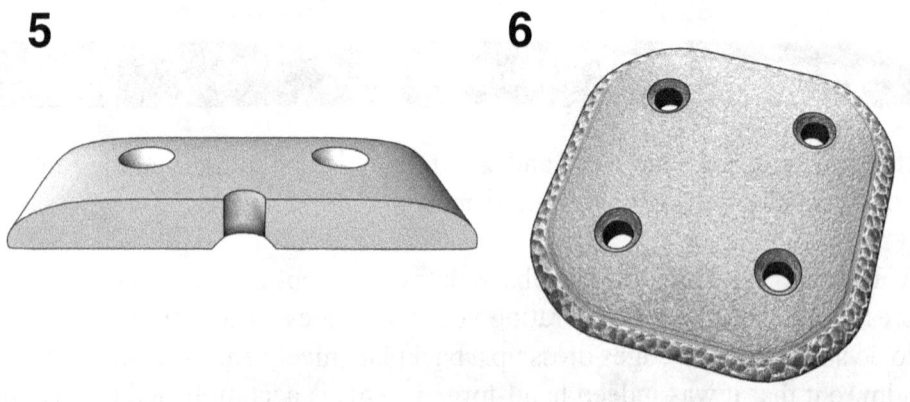

Finished Backplates

Making Hardware with Personality Hook and Eye 189

On the end of the rod, create a circle which will then be tethered to the anchoring eye (Picture 7). A nice, round shape is best achieved by hammering around the tip of the anvil's beak. If the hook and eye set is very small, however, even the very tip of the beak may not be small enough to shape the rod. Consider placing a punch tool firmly grasped in a vise to finish rounding the rod. The finished ring should easily slip onto the eye with room to move freely.

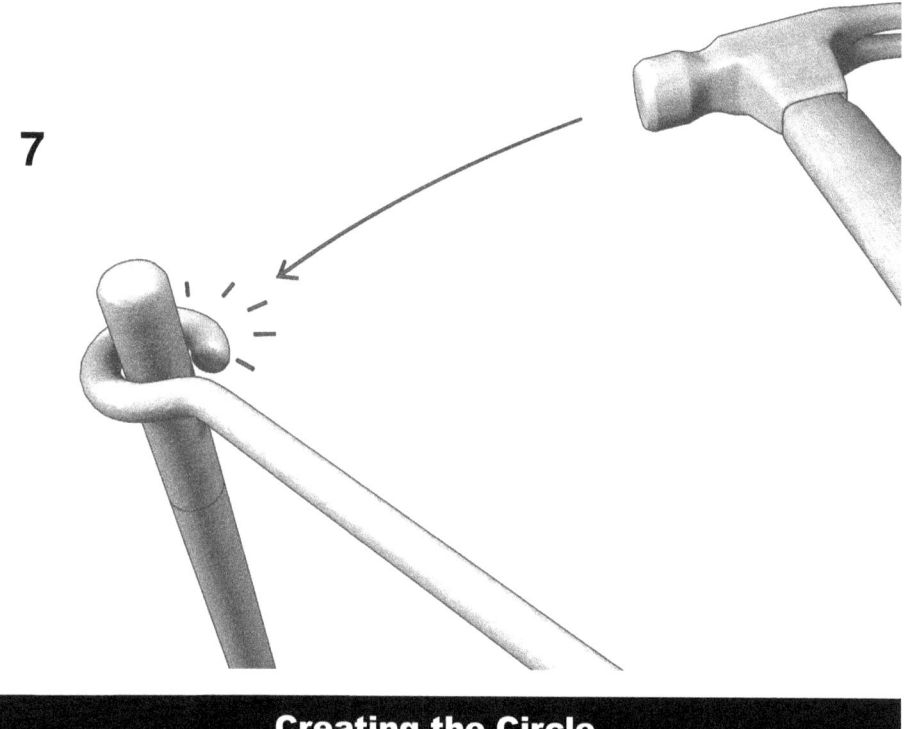

Creating the Circle

The opposing end of the rod will obviously be formed into a hook. The tip of the finished hook should turn up slightly, and this should be done before bending the rod into a hook. Taper the rod to a thin point and then tap it to dull the point (you don't want a sharp needle on the end of your hook). Next, bend it back slightly (Picture 8) and then go ahead and form the hook.

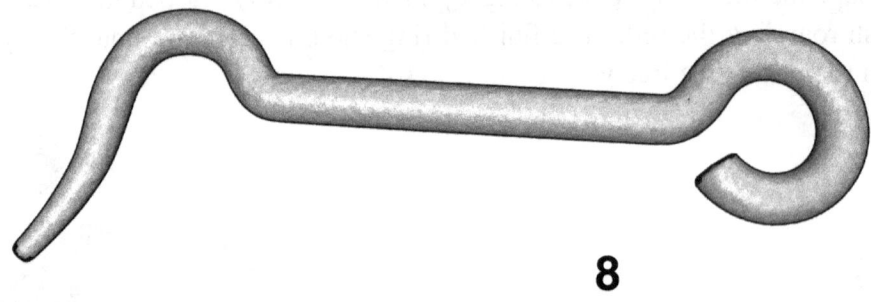

8

Bending

After you have shaped both ends of the rod, you can then thread the ring onto the anchoring eye and rivet it to the backplate. You should fill the rivet the best you can, but don't worry if it is not flush with the backplate. The excess will sink into the wood where the plates are fastened.

You can really spruce up your primitive hook and eye with a twist by using a square rod (Picture 9). Begin with a length of rod that is longer than you intend the hook to be so that you will have something to hold onto while you heat and twist it. If you will recall from Chapter 7, it is recommended that you create the twist in your project before performing other forging techniques. By doing so, if the twist does not come out as you intended and you have to discard the project, you will not have to reforge the ends of the rod.

Once your twist looks the way you want, round the ends as described in Chapter 7. To refresh your memory, hammer the square rod into a hexagon and then continue to hammer until all of the corners are rounded.

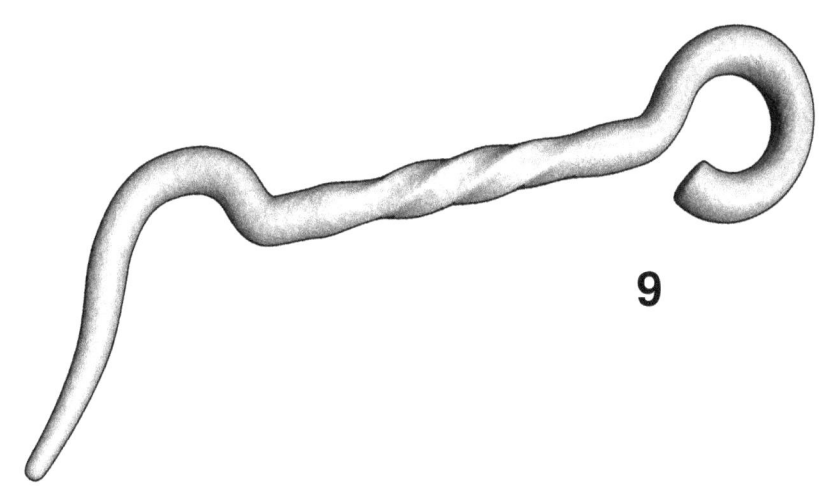

Making a Twist

Note: Take care when creating the ring and hook to ensure they are aligned with each other. If you are holding the piece so that you can look through the ring, the hook should be pointing down, not toward or away from you. It will not function as intended if they are not aligned.

Hasp and Eye

A hasp and eye latch is one you would use if you wanted to secure something with a padlock, such as for a shed or box for valuables. It is crafted in the same way as the hook and eye (well, almost). The only difference would be the placing of the hook onto the flat plate (Picture 10).

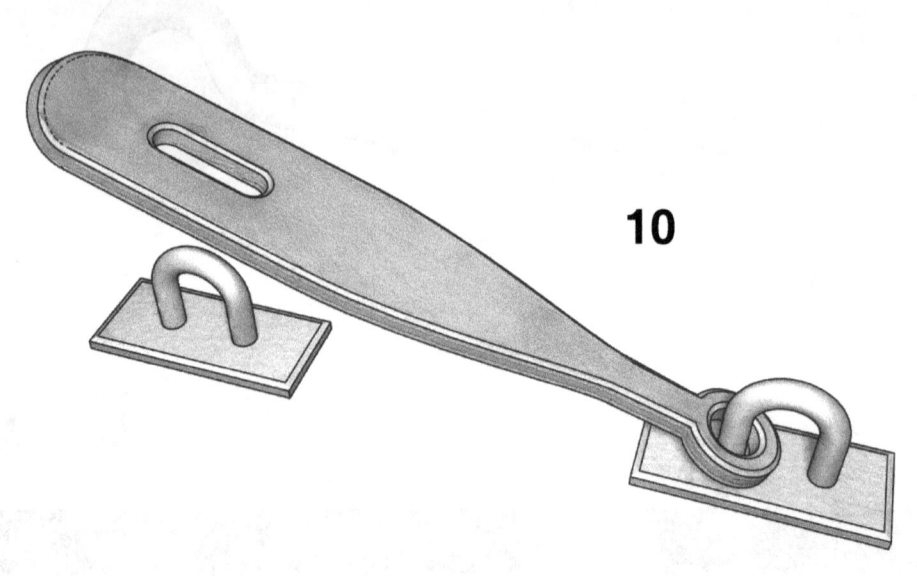

10

Hasp and Eye Latch

Creating a suitable flattened area from a round rod is difficult, unless the hasp you are creating is very small. There would not be enough metal to upset to create the hasp. Therefore, when making larger enclosures, it is best to start with a flat bar. It is much easier to taper one end than it is to try to flatten a rod.

Begin by punching a hole into the flat bar, but be sure to do this a suitable distance from the end of the metal. The hole will end up being oval-shaped, but you will need to begin by creating a round one. There are oval punches that can be used for this purpose, or you can file it into the proper shape. Not surprisingly, the creation of the hole will displace and push the sides of the metal bar outward, which will need to be reshaped through forging. However you create the oval hole, you will need to leave something in place to maintain its shape while you straighten the bulging sides. A tight dowel will do the trick (Picture 11).

Making Hardware with Personality   Hasp and Eye

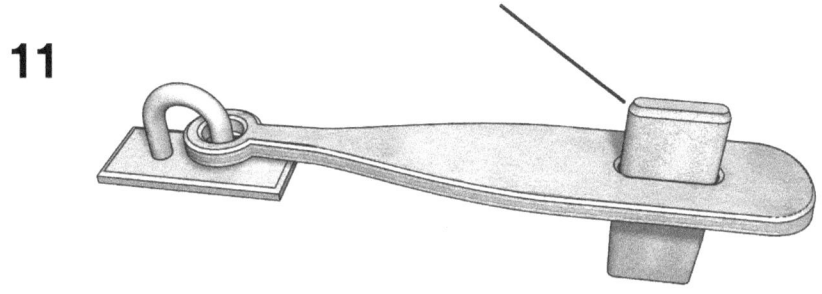

11

Dowel

If you are working on a relatively thin bar of metal, you may need to knock out the dowel, hammer and shape it, and then refit the dowel into the hole for further work. If you find you're really having a hard time straightening the bulges, consider just leaving them there. This will add character to the piece without affecting its functionality. You could call it "decoration". Once you have the hasp the way you want it, draw out the other end of the bar and create a ring for the anchoring eye.

Note: When creating the eye that the padlock will go through, be sure that you arch it sufficiently so that it protrudes far enough through the hasp to allow the placement of the lock (Picture 12).

12

Sufficient Bend

To finish your hasp, consider curving the end slightly. This not only gives your project a bit of flair, but it will also allow for ease in lifting the latch (Picture 13). Although you may want to go crazy with the curve, keep it simple because a harsh curve may interfere with how the padlock lays.

13

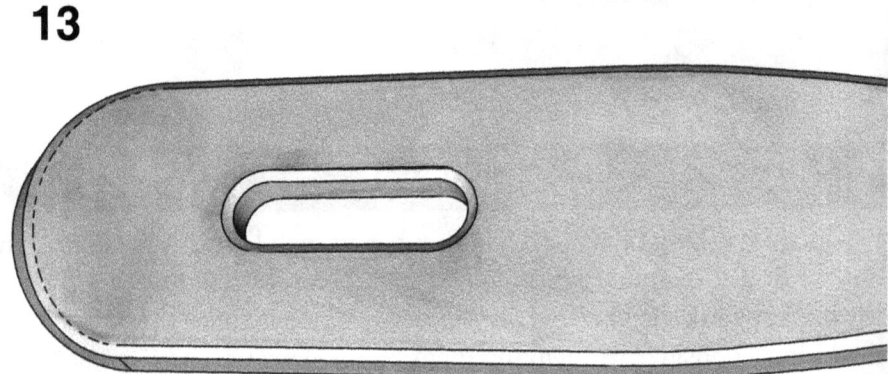

Finishing Your Hasp

If you don't want to bother with the flat hasp, you could create a hasp with a round bar. All you would need to do is form an elongated ring on the end of the bar so that it fits over an eye. It's easy to make and the primitive design fits into the popular "shabby chic" motif.

Sliding Bolt

The sliding bolt is a great choice for a gate, door, or cabinet that can be secured on one side. On a gate or very large door, you can really let your personality and creativity shine.

Making Hardware with Personality Sliding Bolt 195

The basic bolt design involves a rod (either round or square) that is slid into and out of enclosures called *keepers*. The sliding rod includes a handle and a "stopper," which will prevent it from sliding too far in either direction (Picture 14).

14

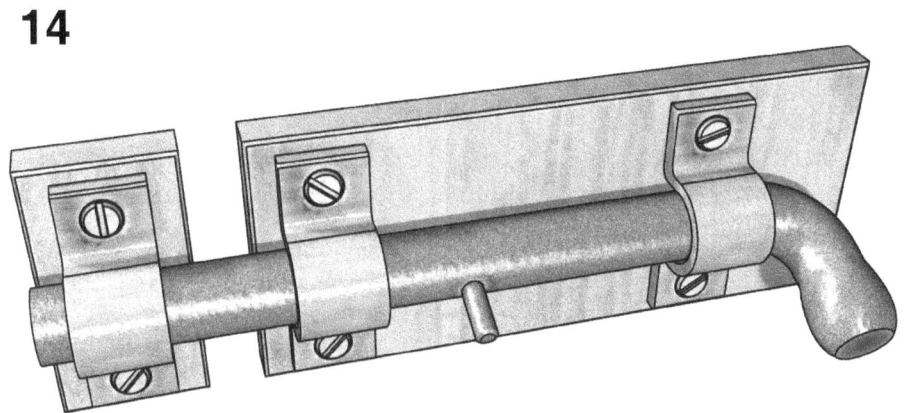

Sliding Bolt

Whether you use a round or square rod depends a bit on personal preference, but you must also consider where the lock will be used and how you want the handle to be positioned. For example, if the lock will be in a high-traffic area, you may want to use a round rod, because the handle of a round rod can be rotated down against the wood when it is not in use, while a square rod cannot.

Determining the sizes of your pieces begins with choosing the rod for the bolt. If the rod is 1/2 inch across (or in diameter), the keeper will need to be made from a 3/4 inch bar and the backplate for the entire apparatus should be at least 1/8 inch thick. If you are making a bolt for a large door or gate, the rod should be 3/4 inch, which would require the keepers to be crafted from 1 inch bars.

After you have chosen the rod for your bolt, you will use it (or at least a piece of it) to shape the keepers.

Open the jaws of your vise to slightly wider than the bolt rod and twice the thickness of the metal bar. Heat the bar to red hot and place it across the vise. Then take the bolt and hammer it on top of the heated bar (Picture 15). Continue hammering the bolt down until it is level with the top of the jaws and the ends of the heated metal can be turned back (Picture 16). If you are forging a thicker bar of metal, this process can be a bit more difficult, but once you get it started, it will take shape more quickly. Consider hammering the heated metal with a rounded punch to start the hollowing process and then continue on as you normally would (Picture 17).

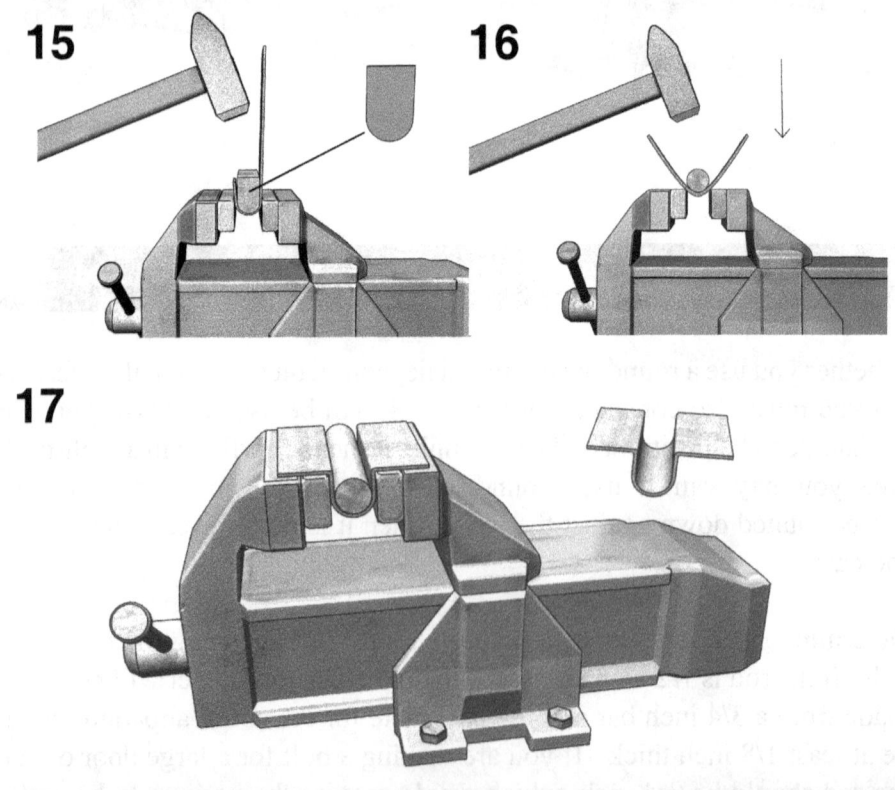

Shaping the Keepers

The process to make a square keeper is the same as that which was just described. However, if the bolt you're crafting is very large, you may have to sharpen the bends of the keepers manually to ensure they will be a good fit. This can effectively be accomplished by truing them around bars of the same size and hammering out the finishing touches (Picture 18). Once the keepers have been formed, you can then trim them down to size, but be sure to make three that match.

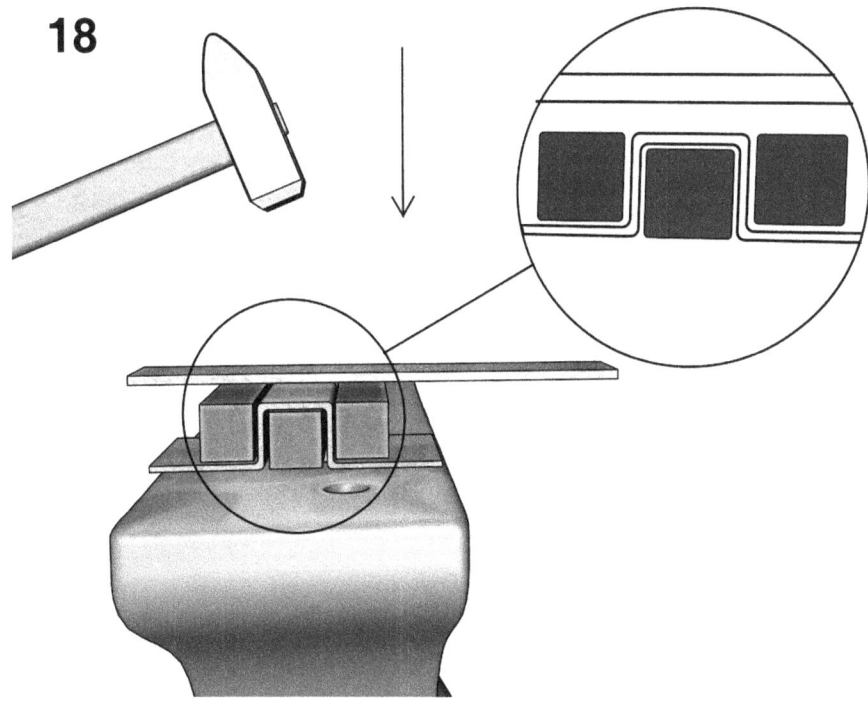

Making a Square Keeper

Next, rivet the keepers to the backplates. This will add strength and stability to the closure, and then drill holes for the screws. However, if the project is small, it is perfectly acceptable to drill through the keepers and backplates together and use the holes to screw the backplate and keepers directly to whatever it is they will be holding closed.

The most basic backplates will be rectangular, and these are certainly the easiest to make. Nothing fancy, but they will most assuredly perform their intended task well. You will now need to cut and shape the bolt.

Obviously, the bolt will need to be long enough to span the area between the two pieces that are being held together. The leading end of the bolt (the end that will go through the two spanning keepers) should be tapered slightly. This will allow the bolt to slip easily into the keeper without getting hung up on the edges, even if the door or gate should shift slightly.

Don't ignore the other end that will become the handle. You still need to bend this end into an angle. The degree of bending is not as important as creating a handle that is comfortable for you. The easiest handle to create and use is one that has a ball on the end. This is accomplished by upsetting the bar on the anvil and rounding it through hammering. If you want to get fancy and create a really unique handle, make sure you do this before you bend the bar. If the bar is bent first, shaping the end while maintaining the bend can be a nightmare. For larger bolts and those that will hold a door closed vertically, the handle will need to be longer and sport a sharper angle. You can really get fancy and let your imagination soar when working with larger bolts (Picture 19).

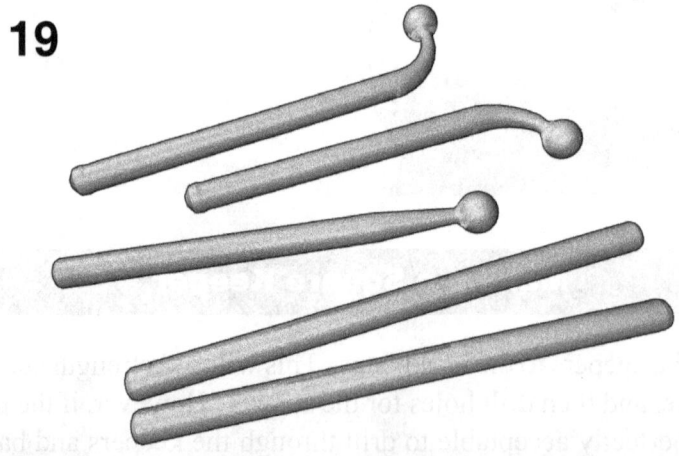

Fancy Looking Bolts

The final step in creating a bolted latch is to drive a stop directly into the bolt. This is really nothing more than a small peg that has been hammered into a predrilled hole. Drill the hole slightly smaller than the peg that will be inserted. Heat the bolt to red, which will expand the metal, and then drive a cold peg into the hole. As the bolt cools, the metal will shrink back into shape, thereby creating a nice, tight fit. For a fit that couldn't possibly come lose, consider driving the peg all the way through the bolt and rivet it in place.

Always check the action of your bolt with the keepers before fastening them to backplates and before mounting. You could even tack them up to double check the alignment and action before securely mounting them.

This simple latch is a great project that will enable you to spread the wings of your imagination. You can hammer the backplates to a beautiful overall finish (Picture 20) and even the keepers can be given a bit of flair (Picture 21). Before mounting, you can use a cold chisel to shape the flat ends of the keepers into a unique shape that is all your own. However you choose to create your bolt, one thing is certain – there will be none other like it anywhere.

20

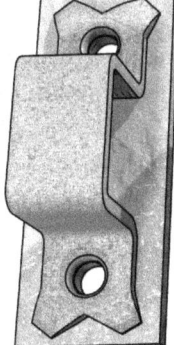

Hammer the Backplates

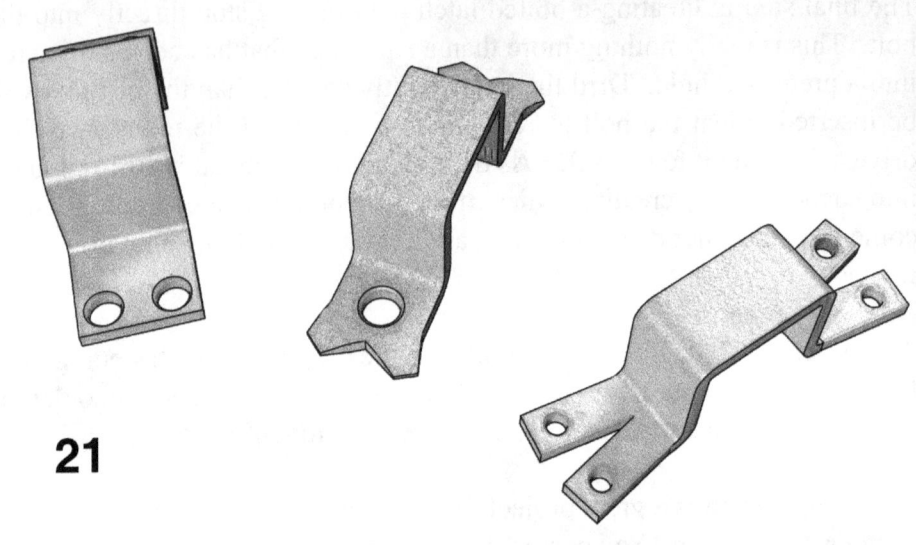

21

Keepers

Hinges

Materials: Metal bar, metal rod, saw, file, mandrel

No matter what catch you choose to place on your project, none is complete without a hinge. Of course, you will want to craft the hinge yourself to retain the uniqueness of your hardware. There are many different types of hinges, and you may even want to study some antique pieces to see how blacksmiths of yesteryear had created them. Most hinges are made with two flat bars of metal encircling a pin while alternate pieces form the knuckles.

Making Hardware with Personality Hinges

The typical modern hinge is comprised of staggered knuckles that are equal in width. It makes sense that there would be an equal number of knuckles on each side of the hinge, as this would allow the two parts to equally share the load of the door. However, when crafting a hinge with only three knuckles, the center knuckle should be wider than those on either side (Picture 22).

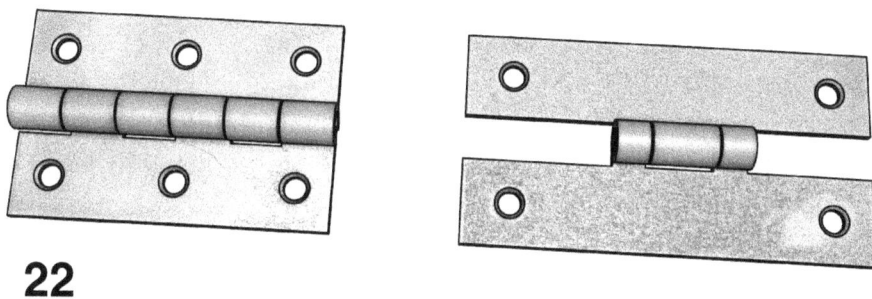

22

Crafting the Hinge with Three Knuckles

A simple hinge is made from a bar of metal that has been wrapped around a pin. For added stability, the wrapped metal can be closed off (Picture 23).

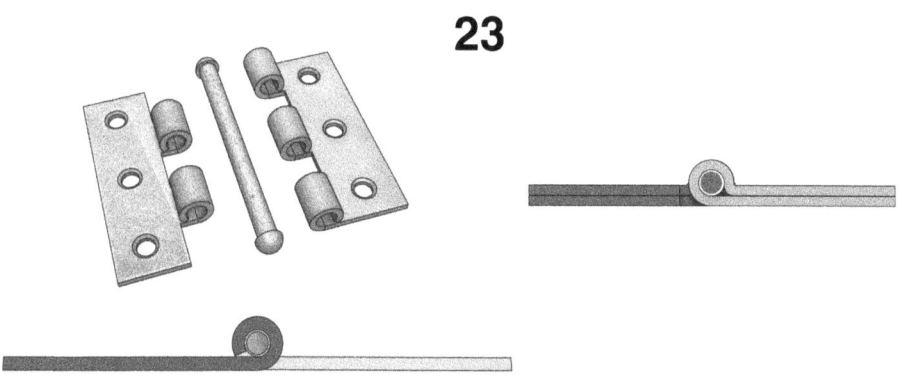

23

Open and Closed Wrapped Metal

First, you'll need to decide how thick the bar of metal and pin should be, which can be a difficult choice. Modern hinge pins are typically no thicker than the metal that surrounds them, but you want to ensure that your pins will be able to carry their share of the load without bending or breaking. A good rule of thumb is to choose a pin that is twice as thick as the metal bar.

Because of the simplicity of its design, a strap hinge is a good first project (Picture 24). The bar should be about 1-1/2 wide x 1/8 inch thick and the pin should be about 1/4 inch in diameter. Do not cut down the length of the bar until after you have formed the knuckles, as this will leave you something to hold.

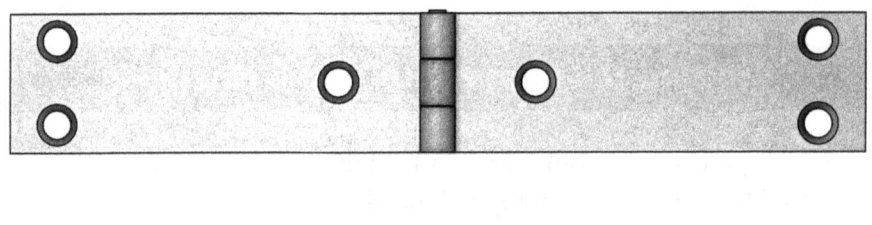

24

Strap Hinge

Begin by bending the red hot metal bar around the edge of the anvil (Picture 25). Once the rounding has begun, turn the bar over and continue closing the curl (Picture 26). Because the pin will be 1/4 inch in diameter, try to estimate and leave enough space for it to slide through.

Making Hardware with Personality   Hinges

25

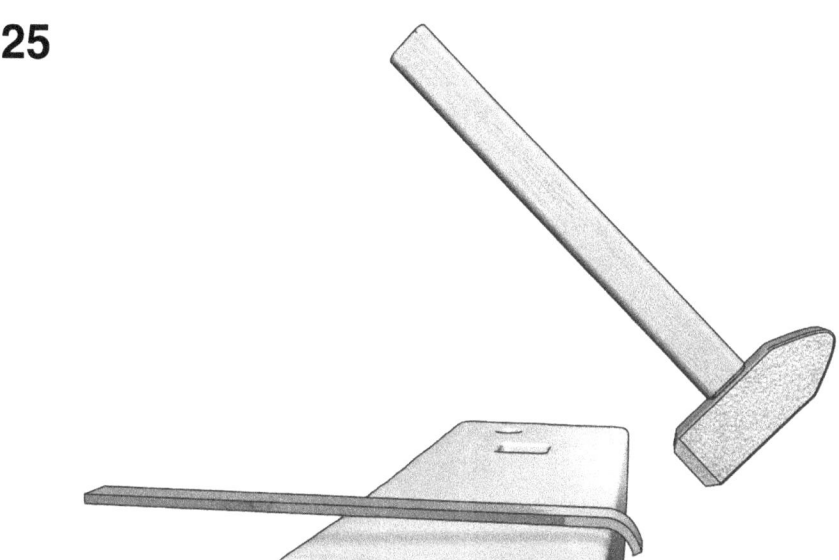

Bending

26

Closing the Curl

Next, take a rod that is the same size as the one you intend to use for the pin, which is referred to as a *mandrel*. This will be used in the final shaping of the curl (Picture 27). You may be tempted to use the pin itself during this process, but it will more than likely be damaged from the hammering. Therefore, it is best to use a mandrel of the same diameter that you can discard or reuse for another project. Also, before shaping the curled metal around the mandrel, you should create a hook on the end. This will make it easier to pull it out after the task is completed. It need not be fancy, just something you can hook your finger through to give it a good pull.

27

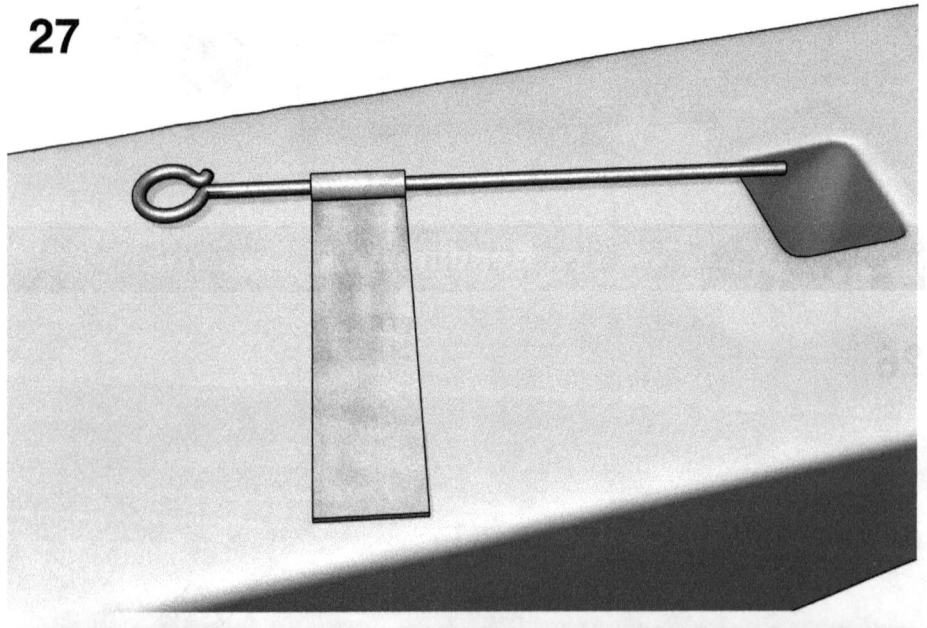

Shaping of the Curl

Continue shaping the curled bar around the rod until you obtain a good fit. If you are not planning to completely close the curl, the edge of a piece of steel will serve well to put a final tightening on it (Picture 28). If you don't have a piece of steel to use, a sharp edge on your anvil will also serve this purpose, but it will be more difficult to control the closing. Pull out the mandrel and repeat the whole process for the second half of the hinge.

Making Hardware with Personality Hinges

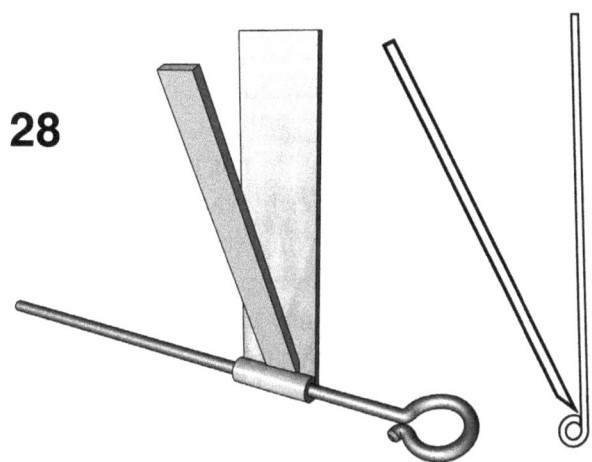

Final Tightening

Once you have the bar curled, measure and mark which parts of the newly formed knuckle will be cut away. Saw down the lines you created, chop out the section with a cold chisel, and then file to shape and smooth the edges. Next, take the second half of the hinge and place it against the finished half. Use the cuts as a guide to mark the second half and create voids to oppose the solid portions of the first half. This will ensure that they will fit together perfectly (Picture 29).

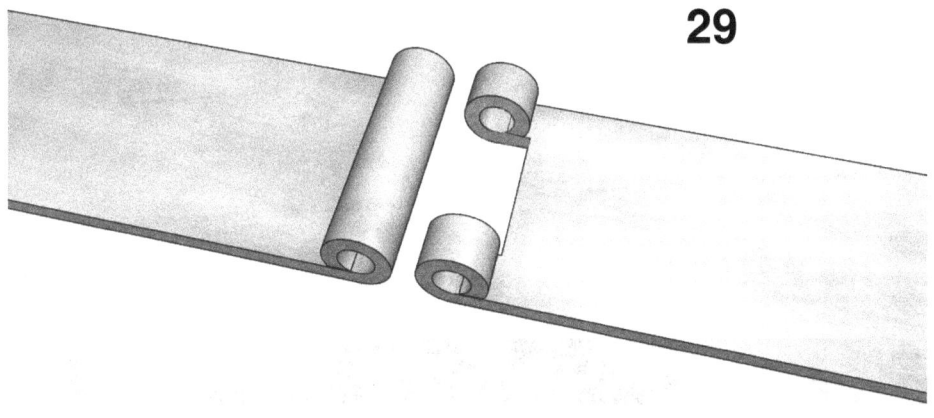

Cutting for a Good Fit

Once you have two pieces that look like they were made for each other (because they were), test their movement with a pin in place. You can then trim the bars down to size and drill screw holes – be sure to countersink.

You can create simple strap hinges that do not look so simple. Add some flair to these typically unassuming hinges by varying their shapes and sizes. Consider crafting any one of the following strap hinges: T-hinge (Picture 30), butt hinge (Picture 31), H-hinge (Picture 32), parliament hinge (Picture 33), or an L-hinge (Picture 34). You may even find that your imagination takes a turn toward the very fancy (Picture 35).

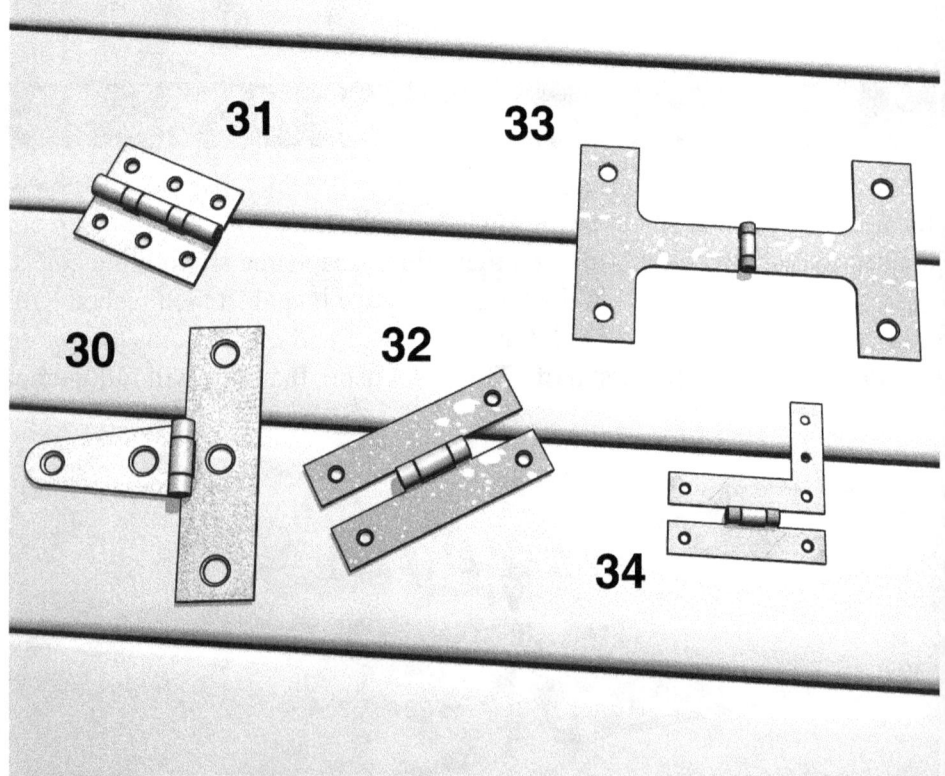

Different Types of Strap Hinges

35

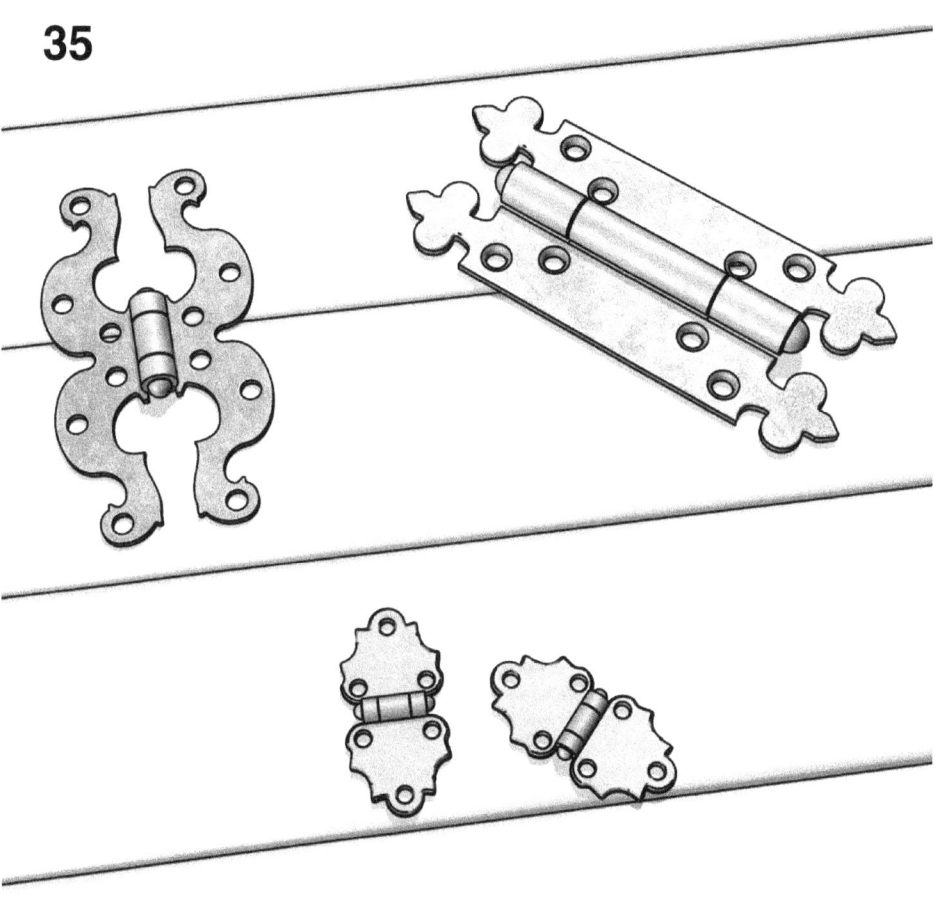

Creating Fancy Knuckles

Mastering the art of creating knuckles takes time and patience. However, once you have the techniques down, you will surely find new and interesting ways to decorate your hinges.

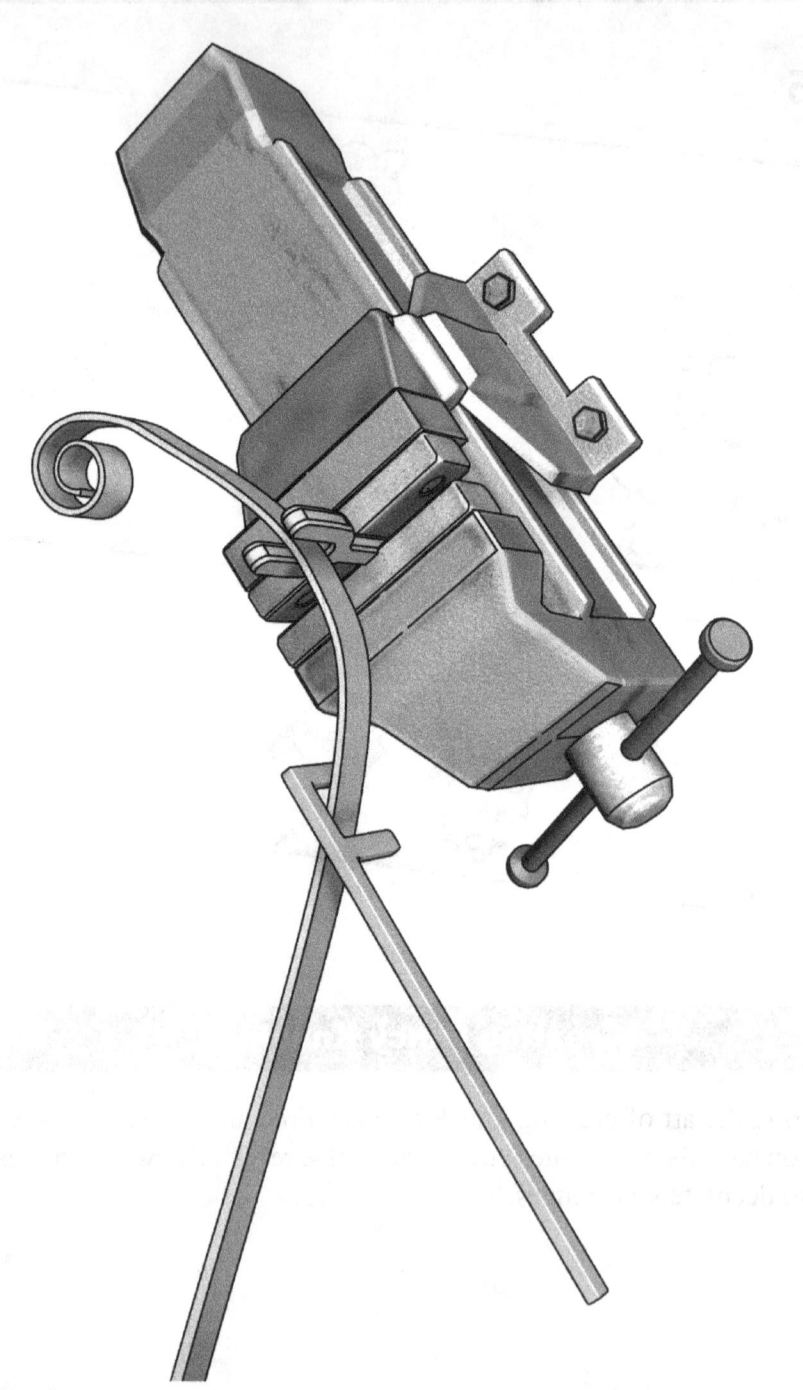

11

Decorative, yet Functional, Smithing

Decorative, yet Functional, Smithing

You can set your imagination free by creating projects that are not only functional, but also have bit of flare. Even the simplest projects can look really impressive. No doubt you will soon be inundated by requests from your family and friends who will want your next one-of-a-kind creation! The projects presented here are simplistic in their design, but with your personal touch, they will come alive and may become some of your most sought-after pieces.

A Place to Hang Your Hat

Material: flat or round bar - size will depend upon the size of the hook to be crafted

You can purchase ready-made metal or plastic wall hooks in many stores. Some attach to the wall with screws, while others adhere to the wall with double-sided tape. Oh, sure, store-bought hooks will hold your personal effects perfectly fine, but do you really want your hat or coat hanging on a characterless hook? Of course not!

A basic wall hook will be a snap to make with the skills you have honed thus far (Picture 1). You will begin with either a flat or round bar of metal, depending upon your preferred method of crafting a hook. Some smiths prefer to begin with a flat bar and then draw the end out into a hook (Picture 2). Conversely, other smiths prefer to begin with a round bar and flatten one end, which will be attached to the wall (Picture 3). Whichever method you choose, creating a ball at the end of the hook will ensure that it will hold a garment without allowing it to slip off.

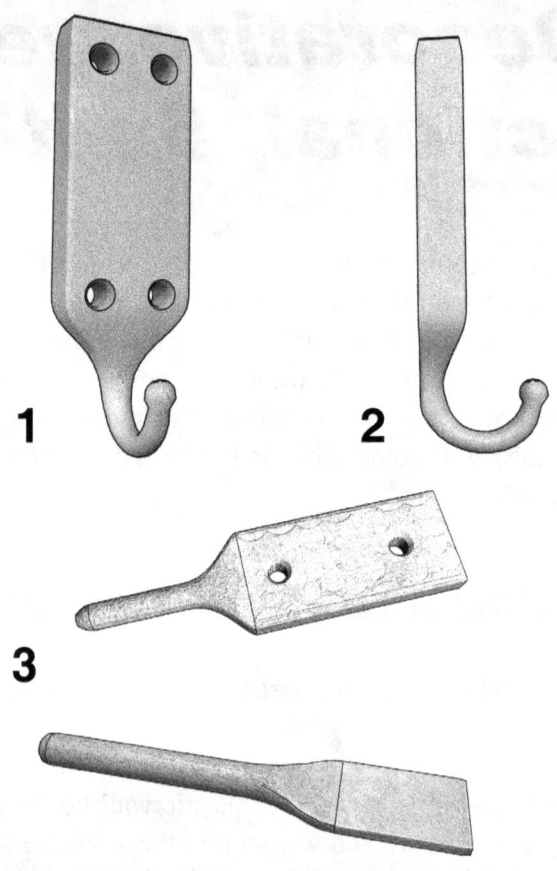

Wall Hook

There are so many ways that this simple hook can be created that it almost boggles the mind. A few tips to enhance your project are offered here, but you will, no doubt, come up with many other ways of making a wall hook into an elegant work of art.

You'll recall in the previous chapter that wall plates were decorated with hammer blows. This same method of adornment can be applied to the back of the hook (Picture 4). To add some simple flair, you can even split the back of the hook (Picture 5). This will not only give your hook a bit of character, but will help to distribute the points of attachment on the back, adding stability and strength to its function.

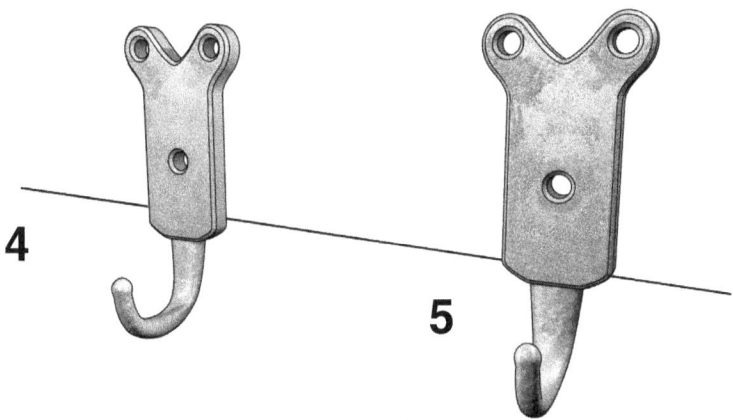

Decoration with the Hammer

There are some more dramatic characteristics you can give your hook, which will also add to its functionality. For example, consider putting a ball on the end of the hook that is not only attractive, but will also better hold a garment left in its charge. This is best accomplished by upsetting the end of the hook before you add the curve. Consider leaving the hammer marks on the ball as this will not only highlight the hand forging, but will also make the ball rough (Picture 6). This will add to the hook's ability to grab a garment so that it does not slip onto the floor.

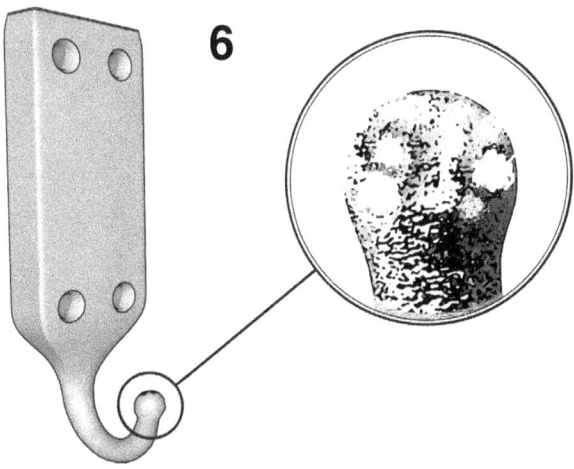

Making a Rough Ball

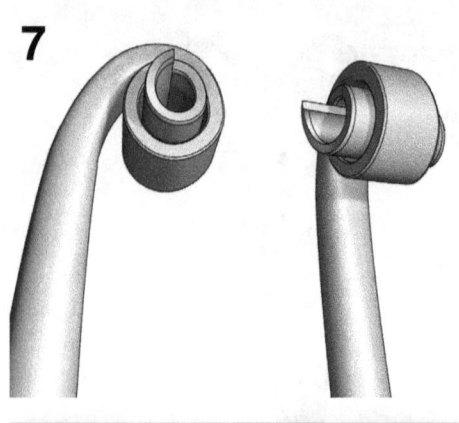

Curling the End

You can adorn your simple hook with some real beauty by adding a curl on the end. It is best to flatten the end of the hook first to give the end some width (Picture 7). This will hold garments more securely and gives the end of the hook a really cool visual effect. First, heat the tip of the hook and then hammer it flat. You may need to heat it again before curling the end to the desired tightness.

If there is an artist's heart beating within your chest, you may want to try adorning your hook with an artistic sculpture. Fancy hooks of yesteryear often sported the heads of animals on their ends (Picture 8). This is not the easiest of tasks, but can be accomplished through careful hammering and the skillful use of punches and files. Consider the creation of a double hook, which will give you two opportunities to create sculpted heads.

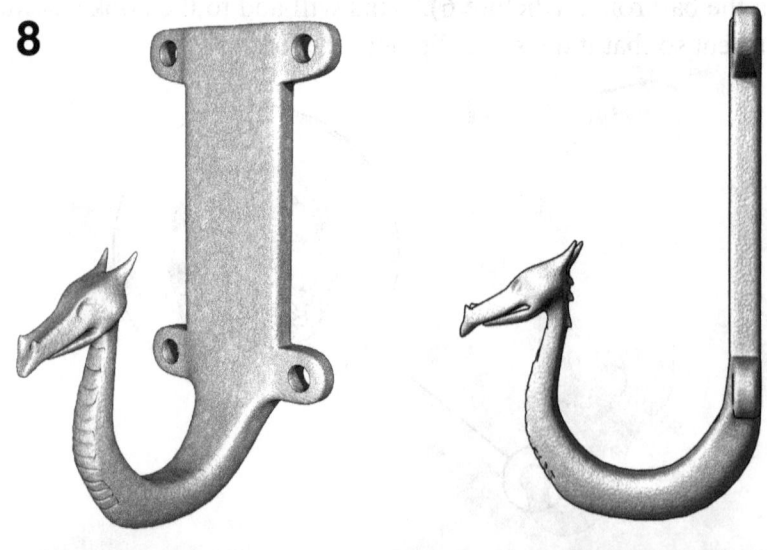

Sculpted Heads

Decorative, yet Functional, Smithing — Wall Brackets

If you do not wish to add any type of ball to the end of your hook, consider forming the hook into what is often referred to as a "swan's neck." This will not only add an air of drama to your piece, but is functional as well (Picture 9).

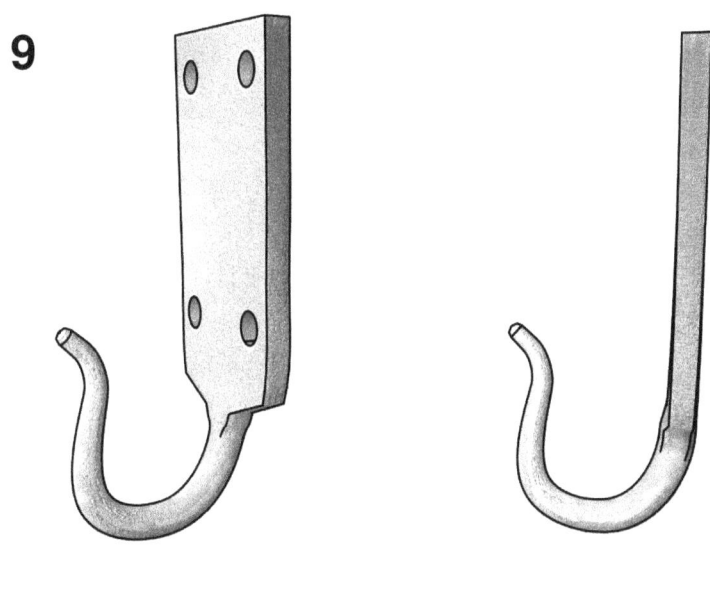

Swan's Neck

Wall Brackets

Material: metal bar - size will depend upon the intended use of the bracket.

Take a metal bar, bend it in half, and then drill some counter-sunk holes for screws (Picture 11). That's all there is to making a wall bracket – it doesn't get much easier than that, does it? Well, that's almost all there is to it. You will, most likely, want to give your wall brackets a bit more character.

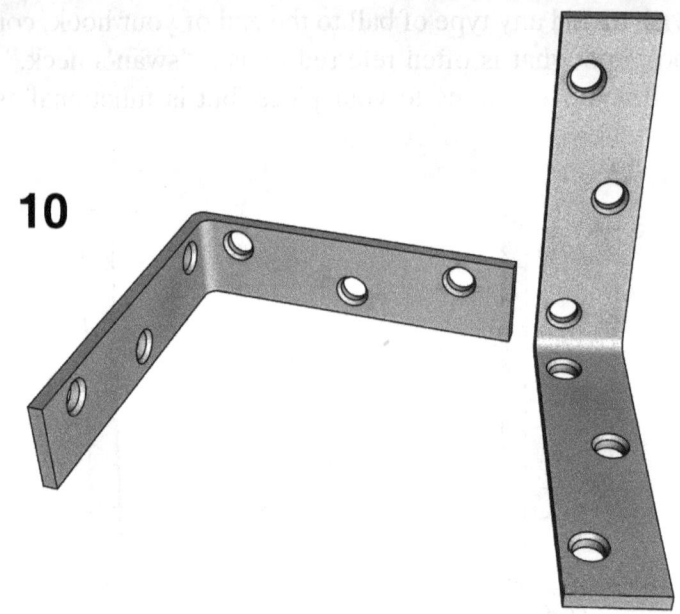

10

Drill Holes for Screws

While it's true that all you need to do is bend a metal bar in half to create a wall bracket, there are some things to consider. As an example, let's say you start with a 1/2 x 3/4 inch bar of metal. Mark the bar where you want the bend to be. You will leave this portion of the bar, plus 1 inch in either direction, at it fullest thickness. When making shelf brackets, be sure that the bend is not less than 90° or the shelf you put on the brackets will look as though it is tilted forward. It is always best to finish a bracket so that it is slightly open when placed against a square (Picture 11).

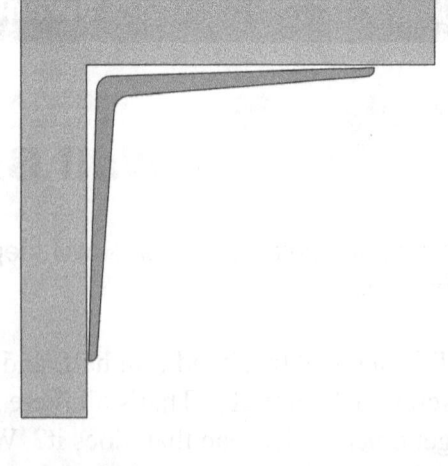

Slightly Open

You can either leave the thickness of the bar as it is, or draw out the ends to approximately a 1/4 inch thick. Consider drawing out the ends to taper out gradually, or shape them into unique shapes that are uniquely your own (Picture 12).

12

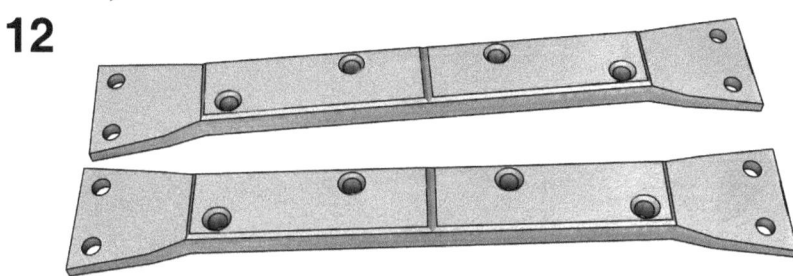

Drawing out the Ends

The previous instructions work great for brackets made from thick metal that will not need any additional bracing. For thinner metals, you will need to add a brace (Picture 13). The bracket is made in the same simple manner of bending a metal bar and then riveting a brace in place. This will not only add stability to the bracket, but is often much more beautiful to look at than a bracket without a brace. Keep in mind that the "leg" of a shelf bracket is typically longer than the top piece; the longer the leg, the longer the brace (Picture 14).

13   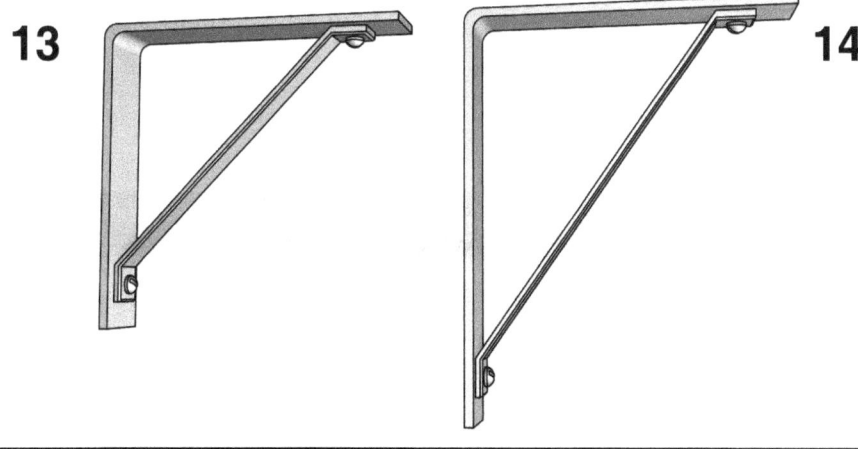   **14**

Adding Brace

You can also allow your imagination to run free with these more interesting brackets. Consider curling the ends of the brace slightly – or dramatically! Curl just the brace, or the bracket, or both (Pictures 15 and 16). Experiment and make your bracket any way that you envision it to be.

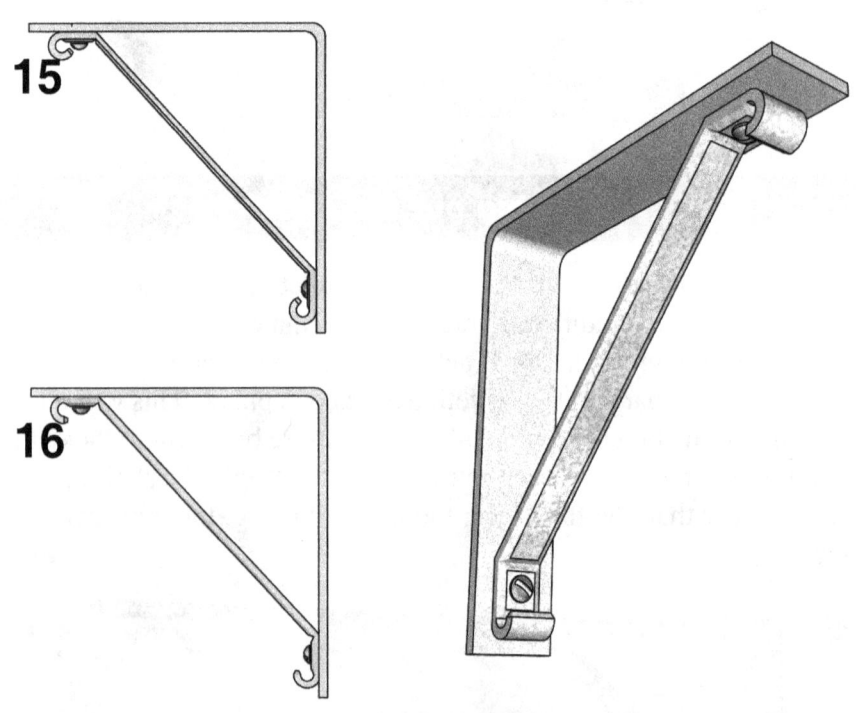

Curling the End

Decorative, yet Functional, Smithing Wall Brackets

For a twist, (pun intended), add a twist to your brace (Picture 17). Keep in mind that as you twist the bar it will become shorter, so this process should be completed first before you measure for placement on the bracket.

A variation of a bracket can be made from three individual metal bars, rather than bending just one piece (Picture 18). This design is ideal for creating a place for hanging your potted friends. The bracket consists of one straight piece with two riveted braces.

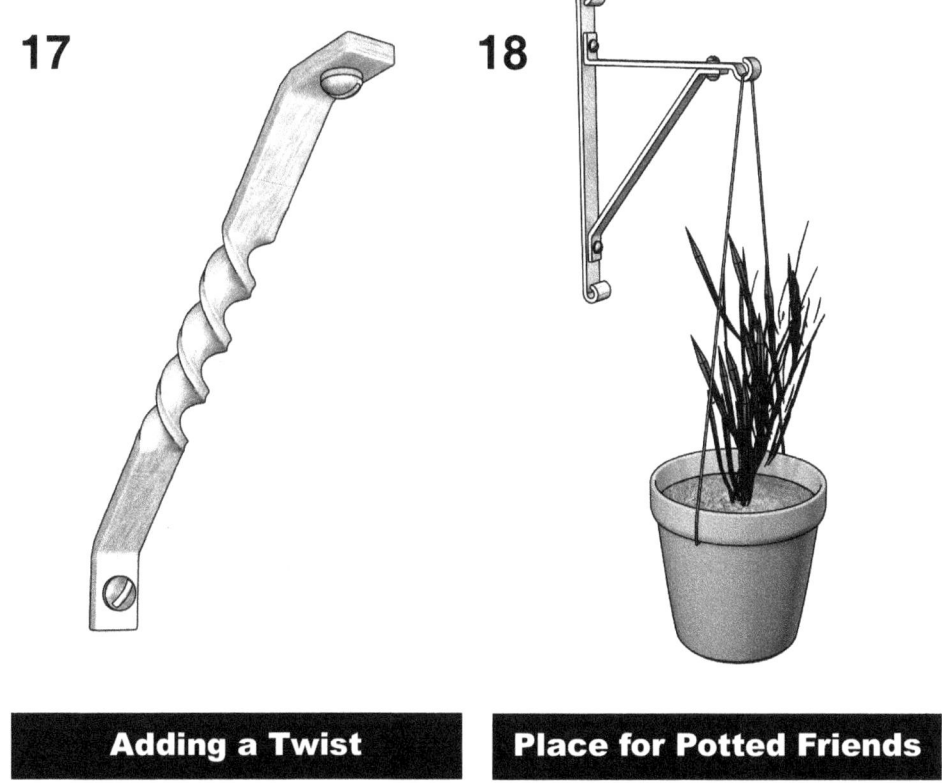

Adding a Twist   **Place for Potted Friends**

Again, your creativity can make the piece as fancy and beautiful as you like. Your angles do not need to be level and sharp when making this design, so you may want to add some curves for a truly dramatic appearance (Picture 19).

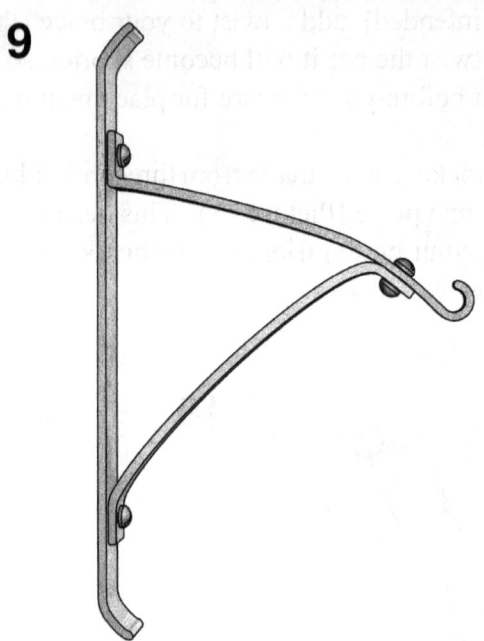

More Dramatic Appearance

Scrolls

Material: flat metal bar that is twice as wide as it is thick, light hammer, various scroll tools (all optional)

Okay, so scrolls are not necessarily functional, but they can be quite amazing. Scrolls are one of the most common adornments seen on the work of blacksmiths over the centuries. They are not very difficult to make, although creating opposing scrolls of the same size and shape takes a great deal of practice and an eye for detail. The easiest type of scroll to create is accomplished with nothing more than a hammer and is typically a very tight curl. There are tools, however, that can help you create scrolls that match each other more perfectly. Let's take a look at the different ways of creating scrolls.

Although both round and square bars can be used to create scrolls, flat bars are the easiest to work with and are recommended while you're learning the technique. First, draw down the edge of the bar to a sharp taper (Picture 20). Try to get the edge as even as possible using a hammer, but feel free to use a flatter to obtain the best edge.

20

Sharp Taper

Next, heat the bar to a yellow heat and begin by hammering the end into a tight curl (Picture 21). The rest of the curling can be accomplished by laying the bar across the face of your anvil and striking the metal with glancing blows, rather than harsh direct hits. The arrows in the illustration show where your blows should be placed. The hammer blows should be progressive as the bar is moved further over the anvil's face (remember your curling lesson)?

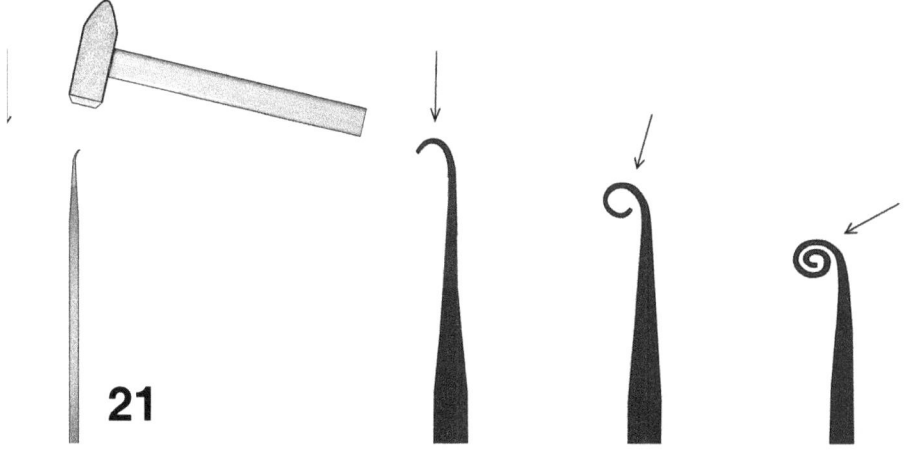

Remember Your Curling Lesson?

You may also need to pull the project back up onto the flat face for hammering to straighten and maintain the thickness of the bar during this process. You will also most likely need to reheat the metal a couple of times before your scroll is complete. Keep in mind that the tip is very thin and will lose heat quickly. By the same token, it will also heat more quickly. Leaving it in the fire too long will result in it melting away.

A truly beautiful scroll is one that is not too tightly wrapped. A scroll always looks best if the tip is tight, but you may want yours to become progressively more open. You can accomplish a wider scroll by placing your blows further back on the bar, while remembering to keep advancing the metal over the anvil (Picture 22). The further back your hammer blow, the wider the curl will be.

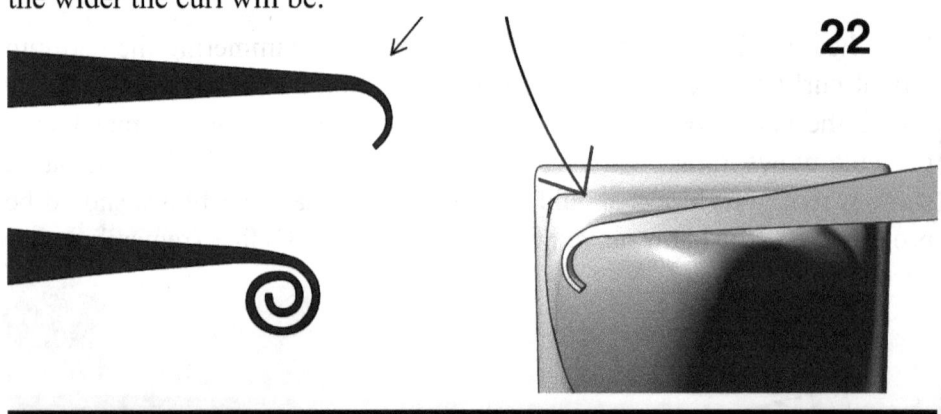

Types of Scrolls

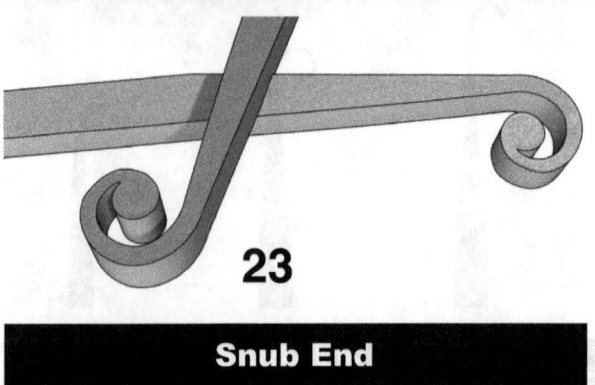

Snub End

In some blacksmith circles, a solid "snub end" tip is the most attractive way to begin a scroll (Picture 23). But, hey, beauty is in the eye of the beholder. The process will be explained here, just in case you like the snub end as well.

Decorative, Yet Functional Smithing Scrolls 223

Begin with a round rod of the desired diameter and cut it almost completely through, leaving a length slightly more than is needed. Start the curve on the bar and then heat both pieces to a welding heat. Weld the pieces together with a mighty welding blow (Picture 24).

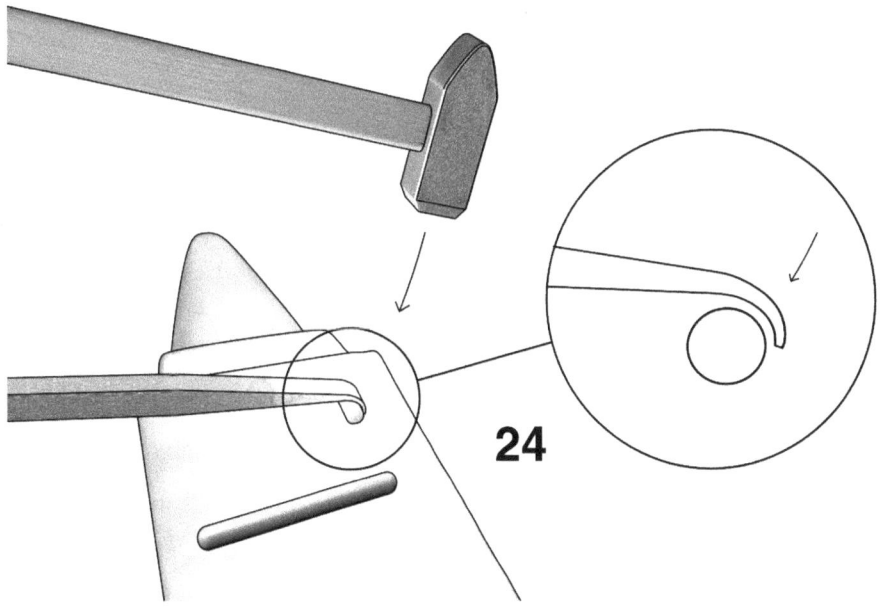

Mighty Welding Blow

Snap off the remaining rod. You may find that you'll need to shape it a bit by hammering or filing. You can then continue creating the scroll as you normally would.

There is not a required number of curls for a scroll. They can simply be an open circle, or can wrap around and around. The real skill of the smith lies in creating matching scrolls with even curls that turn the same and at regular intervals. Depending upon the project you are tackling, this can be a daunting task if numerous scrolls are required. Although machine-like precision is never desired in hand forging, there are some tools that can help keep your scrolls uniform.

A *halfpenny snub end scroll* is a little tool that fits into the hardy hole of your anvil (Picture 25). Its not-so-little name comes from the curve in the side being the same size as an old English halfpenny (pronounced hay-pnee). It is particularly well suited for a bar with a snub end.

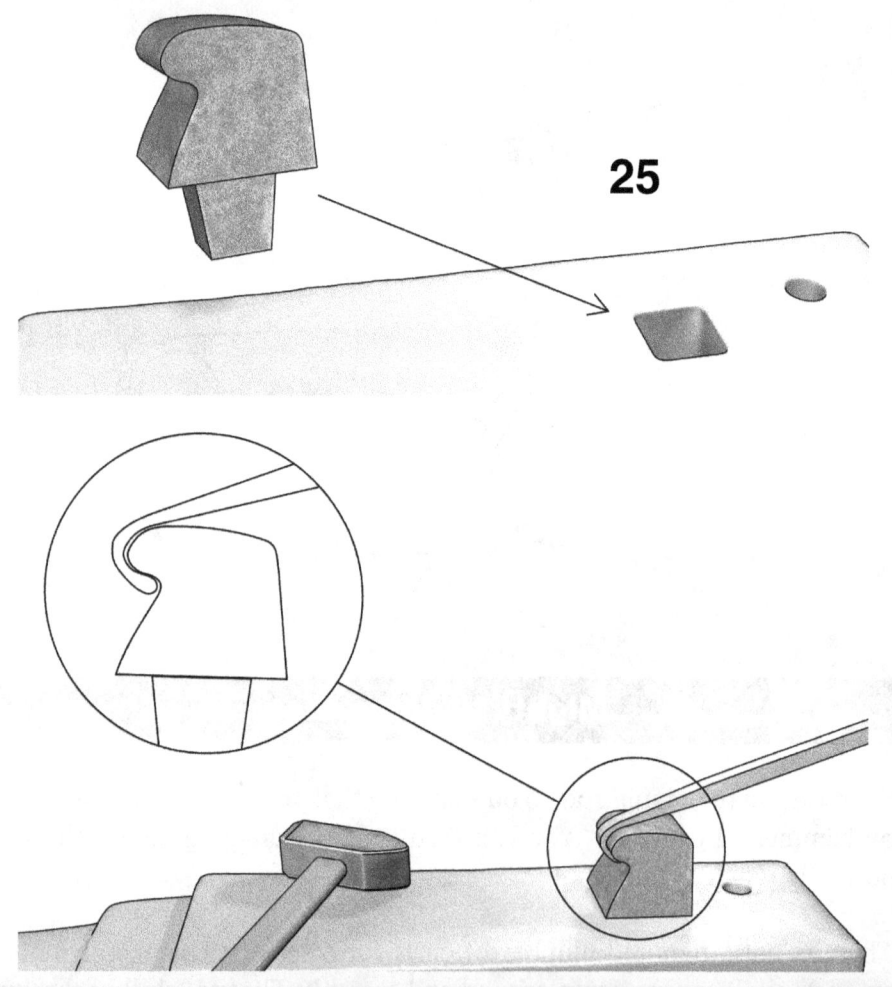

A Halfpenny Snub End Scroll

Another tool that is good to uniformly start repeat scrolls is the *scroll starter* (Picture 26). The tool has an exaggerated bend on the end onto which a curled starter end can be hooked while the hot steel is pulled along the shape of the tool's curve.

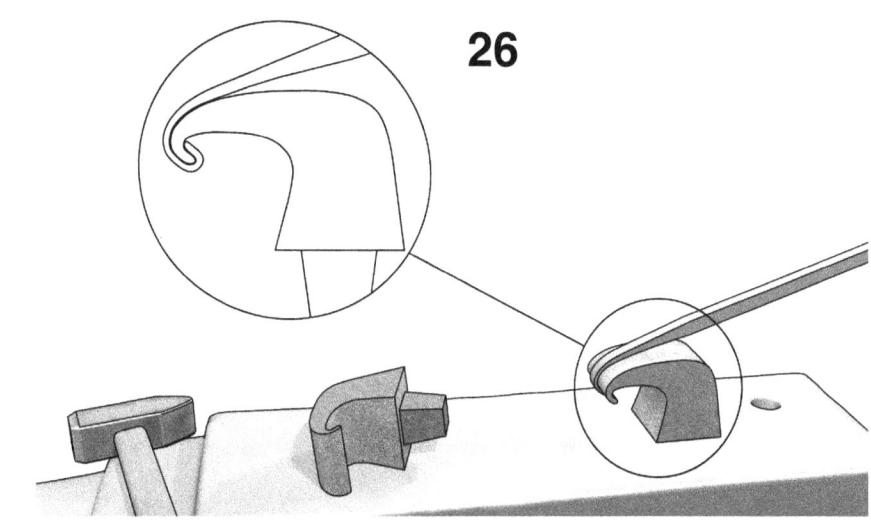

Scroll Starter

If there are many uniform scrolls to be created, you may want to create a sort of template called a *scroll iron* (Picture 27). The template must be made of hardy steel that will not pull out of shape when a hot bar is pulled around it. The center of the template is thinned and has a sharp curl onto which the starter curl can be hooked. This also helps to anchor the hot metal so that it can be quickly wrapped around the template.

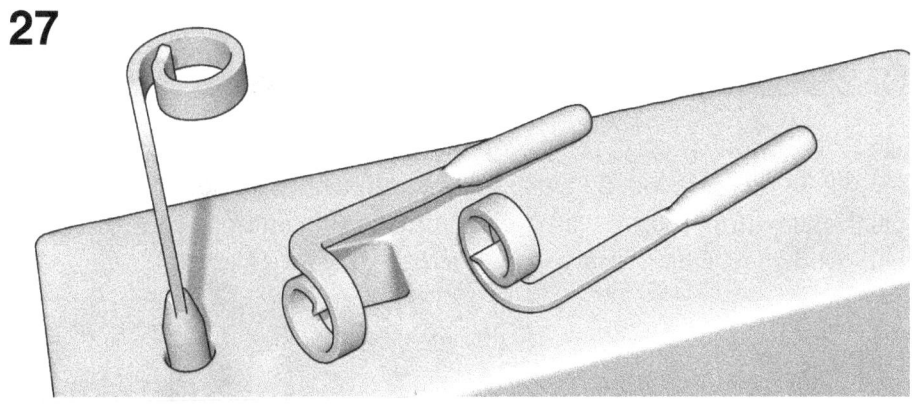

Scroll Iron

You might find that a scroll fork and wrench set works well for you (Picture 28). The fork is held into place by either the hardy hole or a vise. The scroll wrench is used to lever the curves as the hot steel is advanced through. This method allows you to see the curl emerge as you go, allowing for the repair of any mistakes that may occur before moving forward.

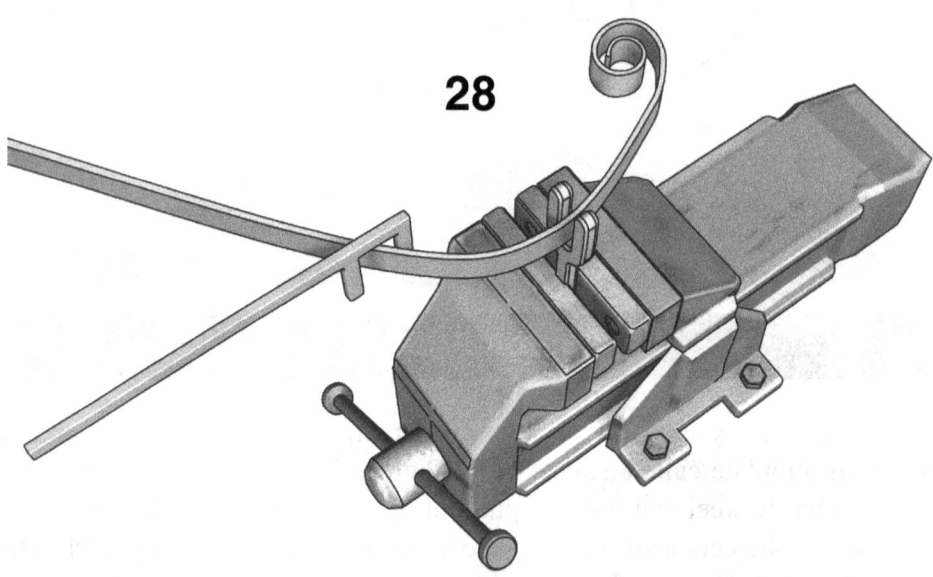

28

Using Scroll Fork and Wrench

Don't be timid when making scrolls. Be bold in your movements! Heat the whole piece of metal so its entire length can be pulled quickly. This is particularly important if you are creating scrolls on both ends of the same bar. And be as creative as you dare to be.

12

Finishing Your Creation

Finishing Your Creation

The creations that you smith are a result of your imagination, sweat, and muscle (and perhaps some blood and a few tears, as well). You will make things that you can leave behind for the enjoyment of future generations. Although they will never meet you, your distant descendents will know something about you – your character, dedication to hard work, and your attention to detail. They can gaze in awe at pieces of hand-forged artistry that have stood the test of time. Art allows us to set our minds adrift to another time and place to wonder about the smith who created such beauty.

Unfortunately, not all forged pieces will last forever. The durability of your creations will depend on the attention you give to finishing them once you are finished forging. The area where you live will also have an effect on your finished projects. Areas of high humidity will cause the degradation of metal more quickly than will dryer climates. Salt air will considerably accelerate the process of decay, so if you live in a coastal community, take the extra time to properly finish your creation.

Recalling the lesson of Chapter 6 and the characteristics of different metals, remember that pieces made from iron will be the most durable. This does not mean, however, that iron is immune to rust and decay. If left unprotected, rest assured that the iron will corrode eventually. Sometimes the first layer of rust that forms on iron will actually serve as a sealant against further decay, but it is best not to rely on letting the rust do the work for you. In all likelihood, the rust will continue to spread like a cancer, deeper into the metal, destroying all of your hard work. All alloys that contain iron can and will rust if not protected. Mild steel, which is the metal most modern smiths use, is particularly prone to decay if left unprotected. The rust will form and spread on the mild steel rather quickly, effectively altering the metal and ruining your creation.

There are alloys made today where a combination of metals is added to help protect the finish from decay. Some will have the characteristics of iron, where a thin layer of rust is all that may form. Other alloys produce stainless steel, which is quite resistant to rust. Logic would suggest that these metals would be great to use in creating your projects, but they are not suitable for forging. They will not respond properly to the heat treatments applied in the smith shop. They can still be heat treated, but they require elaborate processes with equipment that is not found in the typical smith shop.

Removing Scale

Scales will naturally form as you heat and forge your creations. To properly protect the metal for years to come, the scales must be removed and there are several ways this can be accomplished. Heating the metal to a high temperature will burn off the scales, but this is probably not a great method for a finished project. Although it requires a bit more work, knocking the steel against the anvil during the last heating will remove some of the scale, but you will need to complete the task with a wire brush. This can be done manually or with a powered wire brush. As you apply this process, you will note the metal taking on a burnishing effect. This is normal and the color can be restored to an even tone with an oil or wax process, which will be described later. If your creation is very intricate, or has small crevices in which scales are formed, soaking it in a hot brine will help to loosen any scales that may not be visible or accessible with a wire brush.

To truly remove all of the scales from your project, you will need to subject it to an acid bath. The explanation of this process will be prefaced with some words of caution, as all acids, even if diluted, are very dangerous. Acids will eat through whatever they touch, including the container that holds them. Glazed earthenware with a wooden lid is the best choice for an acid container. Protect your eyes and skin at all times! Even if you are well protected, accidents still can occur. Before you begin, have an alkali mixture ready just in case you do get acid on your skin.

Baking soda and ammonium hydrate are both good alkali choices and will neutralize acid quickly. Follow the administration of an alkali with copious amounts of water.

You will need to prepare a pickling bath of acid and water. You must always pour the acid into the water – **NEVER POUR THE WATER INTO THE ACID!** If you pour the water into the acid there is a greater chance that the acid will splash up and hit you! An effective pickling bath is created by pouring one part sulfuric acid into ten parts water, although phosphoric acid in the same proportion may also be used. For a really potent bath, add one part sulfuric acid and one part hydrochloric acid to sixteen parts water.

There is not really a set amount of time to leave your creation in the toxic bath. You will need to remove it periodically for examination to see if the scales are gone. Keep the dripping project away from your face and body during inspection and, although this may be stating the obvious, – **USE TONGS!** Never stick your hand in the bath or handle the wet metal! After each use of the tongs, wash them off with water to ensure that none of the bath remains on them. Once you are satisfied that your creation is scale-free, wash it off with water, followed by a treatment with an alkali mixture, then repeat washing with water.

Protecting Your Creations

For a more gentle treatment to protect your creation, rub the creation with a cloth soaked in linseed oil, then return it to the fire to warm it (sounds kind of nice, doesn't it)? The heat will cause the oil to carbonize; you can then remove it from the fire, rub it again with the linseed oil, and leave it to cool. Once the piece is cooled, wipe off the excess oil. Although this process will protect your creation from rust for a long time, it should nevertheless be repeated occasionally. You can also get the same effective protection by using beeswax instead of linseed oil.

For super protection, consider using both beeswax *and* linseed oil. Your creation is sure to appreciate your extra effort. Warm about one quart of linseed oil and then flake about six cubic inches of wax into the oil with a knife. Stir the mixture until all of the wax is melted and then apply it in the same way you would if you were using just the oil. Beeswax and linseed oil are not exclusive ingredients for this process. You can experiment with your own combination of oils and waxes.

There are also rust-inhibiting fluids that can typically be purchased from a store that sells automobile supplies. They work by forming a layer of their own corrosion that acts to seal the surface of the metal. It will also stop any further decay to metal that has already seen the effects of rust.

Painting

Be bold – go ahead and paint your creation. This is an especially good idea if the project you have created will be outdoors. The paint will protect the metal well, but is not everlasting. Paint will eventually peel and the metal will need to be repainted. It is also very important to properly prepare the metal for painting or the paint will end up cracking and will peel much sooner that you expected.

It is best to paint your creation soon after the treatment to remove scales has been completed. It is also essential that a layer of steel primer be applied before painting the metal. Painting directly on the wood or over primer that was meant for wood and other materials, will result in the paint peeling away. Steel primer has a special action that etches the metal, allowing it to better adhere. Next, apply a layer of matte paint (you may need to apply a second coat if the first does not give good coverage over the primer). Finish your piece with a glossy paint. This will not only add beauty, but the material that makes the paint shiny also makes it tough.

Cool Treatment

There is a really cool process you can use to add flare to your project called *patination*. You will recall from Chapter 7 and the lessons on heat treatments that polished steel that has been given the oxide colors test is less likely to corrode than untreated steel, so this process has an added benefit. This also applies to both mild steel, which cannot be tempered, but can still produce oxide colors, and to tool steel. Oxide colors can only be produced on smooth surfaces that can be polished to a high shine. However, it is also possible to produce them on an uneven surface. Wire brush the raised areas, leaving the recesses black, and hold the project over a single gas flame rather than a fire. You will see the colors emerge and you can stop at the color you wish. You may even want to have graduated colors spread out across your creation. Once you have the look you want, remove the flame and rub the metal with oil or wax (or both). Do not return the metal to the fire or you will lose the color.

Note: Not every piece you make will need to have these extra processes. In particular, the tools you use most often will resist corrosion simply because you use them a lot. For those tools you do not use very often, rub them with some lanolin before storing them.

In Closing

The journey from a beginner smith to an experienced craftsman (or craftwoman) is one wrought with trials and errors. What an exciting time it is! Learning all of the processes that will allow you to make the most beautiful creations and hone your artistic skills was not easy, but we hope that it was, and will continue to be, enjoyable. Make no mistake about it – you are an artist! Go forth from this day forward and fill the world with your artistry!

Glossary

Abrasive cut-off wheel - an abrasive wheel that is used to cut hardened steel

Alloy steel - a metal that contains a mixture of two or more metals

Annealing - the process of softening high-carbon and mild steel by a process of heating and slowly cooling (also referred to as *normalizing*)

Anvil - a large, heavy steel tool on which a smith will hammer and shape metal

Ball peen hammer - a general purpose hammer that has one flattened end and an opposing round end

Beak - the end of an anvil that is used to round metal – also called the horn

Bellows - a device used to blow air into a fire to increase the intensity of the heat

Blacksmith - a smith whose specialty is working with iron

Body - main part of the anvil

Borax - a chemical that is used to help fuse together metal parts that are being welded together

Brine - a quenching bath consisting of water that is saturated with salt

Buffing compound - a mixture used in conjunction with a cotton buffer to polish metal to a high shine

Carburizing - a process that adds a thin layer of high-carbon steel to mild steel, also called case-hardening

Glossary

Case-hardening - a process that adds a thin layer of high-carbon steel to mild steel, also called carburizing

Chainsmith - a smith who specializes in creating links for a chain

Coal box - the container where a blacksmith's coal is stored

Coke - a by-product that results as coal is burned in a fire

Cotton buffer - tool used to polish finished metal work

Cutting steel - the process of cutting steel

Drawing out - the process of thinning and lengthening metal

Drill press - a machine that uses drill bits to drill holes

Ductile - malleable, able to be drawn out while maintaining the strength and integrity of the metal

Electric driven fan - an electronic device used to blow air into a fire to increase the intensity of the heat

Face - main working surface of the anvil

Farrier - a smith who specializes in creating horseshoes

Ferrule - a metal band around the top of a tool handle that adds stability and prevents a wood handle from splitting

Fireclay - a type of clay that will not crack when placed in a fire for long periods of time

Flatter - a handled tool with a broad, flat head that is placed over hot metal and hammered to create a flat surface on the metal

Flux - clean, fine sand, or a combination of sand and borax that is added to the ends of pieces of metal to help the ends bond together during welding

Forge - contains the fire where the metal is heated

Fuller, bottom - a tool seated in the hardy hole that aids in drawing out larger areas of metal

Fuller, top - a tool used in conjunction with a bottom fuller, but is placed over heated metal to aid in drawing out larger areas of metal

Grinding wheel - machine with a spinning stone wheel used to shape metal

Hand-cranked blower - a crank driven device used to blow air into a fire to increase the intensity of the heat

Hardie hole - hole in the heel of the anvil that accommodates a hardie tool and various other tools

Hardie tool - any tool with a square base that is designed to fit into the hardie hole of an anvil

Heading tool - a tool with a hole to accommodate the diameter of a steel rod, which is inserted and upset on the end to create a head for a bolt

Heat treatment - the heating and cooling of metal to change its characteristics, which includes annealing, hardening, normalizing, and tempering

Heel - the back end of an anvil through which holes are typically drilled

Horn - end of the anvil used to round metal – also called the beak

Jumping up - the process of shortening and thickening metal, also called upsetting

Kevlar - a fire-resistant material which can be used to make protective clothing

Glossary

Mandrel - an iron block or round rod around which metal parts are shaped

Nailsmith - a smith who specializes in making nails

Normalizing - the process of removing stress points from high-carbon and mild steel by a process of heating and cooling slowly, also referred to as annealing

Oxidation - the color spectrum that appears on polished metal as it is heated, which is used as a guide for the desired tempering heats

Pickling - the process of removing scale from metal by soaking it in a chemical bath

Poker - a tool used to push coke and coal into the fire

Primer - a base coat that is applied to metal before painting so that the paint will better adhere to the metal

Pritchel hole - hole in the heel of the anvil that provides clearance when punching through hot steel – also called the punch hole

Punch hole - hole in the heel of the anvil that provides clearance when punching through hot steel – also called the pritchel hole

Quench - the process of cooling hot steel in liquid

Quenching bath - a container of liquid, typically water, that is placed near the forge for quickly cooling heated metal

Rake - a tool used to pull coke and coal into the fire

Rivet set - two-part tool with hollows used for creating round-head rivets

Scarfing - upsetting and tapering metal to thick, tapered ends in preparation for welding

Screwing tap wrench - a piece of metal with a square hole near its center

Scroll - decorative, curling metal

Scroll fork - a tool that resembles a two-pronged fork that is used to pull heated metal to create a curve

Scroll iron - a template made from iron around which a metal bar is pulled around to create a scroll

Set - a tool resembling a hammer that is used to cut and shape metal

Smoke catcher - "catches" the smoke from the forge's fire and feeds it back into the air inlet, thus keeping the fire relatively smoke-free

Spatula - a tool used to lift and move coke and coal around the fire

Swag set - a tool consisting of a top and bottom that are grooved and used to shape round pieces of metal

Swag spring - top and bottom swag tools that are connected with a spring handle

Table - step down area from main surface of the anvil, the metal is softer so as not to damage the cutting edges of tools

Tang - the end of a tool that is pointed and driven into a handle

Tempering - a process to reduce the hardness and brittleness of fully hardened metal

Tripoli - coarse buffing compound used in conjunction with a cotton buffer to finish metal to a high sheen

Upsetting - the process of shortening and thickening metal, also called jumping up

Vise - A machine with two jaws that are closed with a tightening screw to

Glossary

Waist - the narrow section of an anvil

Welding - the fusing together of two pieces of metal using heat

Whitesmith - a smith whose specialty is working with lead

standard publications, inc.

VISUAL GUIDE TO LOCK PICKING

Order Form

Cost of only **$24⁹⁵**
(plus 7.5% sales tax in IL)

Use this easy order form to order by mail. Include $3 extra if you want us to send it Priority.

Name	
Email	
Address	
City, State ZIP	

Credit Card Number	
Exipiration Date	
Signature	

Please Mail to: **Standard Publications, Inc.**
PO Box 2226
Champaign, IL 61825

www.ingramcontent.com/pod-product-compliance
Lightning Source LLC
Chambersburg PA
CBHW071708160426
43195CB00012B/1623